D1236619

FROM
MINERALOGY
TO GEOLOGY

Science and Its Conceptual Foundations
David L. Hull, Editor

FROM
MINERALOGY
TO GEOLOGY

The Foundations of a Science, 1650–1830

Rachel Laudan

The University of Chicago Press
Chicago and London

Rachel Laudan teaches history of science and technology at the University of Hawaii at Manoa. She is the editor of *The Nature of Technological Knowledge: Are Models of Scientific Change Relevant?*

The University of Chicago Press, Chicago 60637
The University of Chicago Press, Ltd., London

© 1987 by The University of Chicago
All rights reserved. Published 1987
Printed in the United States of America

96 95 94 93 92 91 90 89 88 87   54321

Library of Congress Cataloging-in-Publication Data
Laudan, Rachel, 1944—
    From mineralogy to geology.

    (Science and its conceptual foundations)
    Bibliography: p.
    Includes index.
    1. Earth sciences—History. I. Title.
II Series.
QE11.L35 1987   550'.9   86-30783
ISBN 0-226-46950-6

For Larry, who has
always paid me the compliment
of being my severest critic

Contents

Figures

Tables

Acknowledgments

In the last couple of decades the history of geology, while not enjoying the explosive growth that has characterized many other areas of the history of science, has nonetheless flourished. In the period covered by this book—namely the seventeenth, eighteenth, and early nineteenth centuries—one thinks particularly of the work of the Carozzis, François Ellenberger, Rhoda Rappaport, and Kenneth Taylor on the French, Walter Cannon, Roy Porter, Martin Rudwick, Nicholaas Rupke, and Hugh Torrens on the English, Tore Frängsmayr on the Swedes, Nicoletta Morello on the Italians, and Martin Guntau and a number of other East German scholars on the Germans. Martin Rudwick has illuminated the study of paleontology, Gordon Davies of geomorphology, Gabriel Gohau of mountain formation, David Oldroyd of mineralogy and Seymour Mauskopf of crystallography. Our knowledge of Buffon, Hutton, and Werner has been much improved thanks to the labors of Jacques Roger, Arthur Donovan, and Alexander Ospovat, respectively. Numbers of other scholars have made more limited, but nonetheless valuable, contributions to specific topics bearing on the history of geology. This activity has drawn attention to topics in the history of geology that had hitherto been largely neglected.

I depend heavily on the work of these scholars, as must any author who has the temerity to tackle two-and-a-half centuries of a discipline's development. My interpretation and synthesis of the theses advanced in numerous articles and monographs may well differ from anything their authors had intended. Yet the history of geology in this period still seems to me to be as much in need of a general, albeit tentative, thematic treatment as it was in the early 1960s (Rappaport 1964; Eyles 1966). I shall have achieved my purpose if I draw attention to the new themes occupying historians of geology, even if my synthesis stimulates as much controversy as agreement.

My colleagues in the history of science have been most generous with their knowledge and their time, particularly Tore Frängsmayr, Gabriel Gohau, Janis Langins, Homer LeGrand, Jim Lennox, Sy Mauskopf, Sally Newcomb, MaryJo Nye, and Jim Secord. In thinking about the conceptual foundations of geology, I have benefited from conversations carried on over a number of years with David Kitts. I regret that the still significant effects of national boundaries on the transmission of knowledge have prevented me making proper use of *Theorie der Geowissenschaft* by Wolf von Engelhardt and Jörg Zimmermann. I have benefited from detailed reactions to drafts of all or part of the text from a number of colleagues, including Stephen Brush, Albert and Marguerite Carozzi, Arthur Donovan, Norman Gilinsky, Mott Greene, Anthony Hallam, Jon Hodge, Larry Laudan, Homer LeGrand, David Oldroyd, Alexander Ospovat, Rhoda Rappaport, Jacques Roger, Ken Taylor, and Hugh Torrens. David Hull deserves special thanks for giving me the courage of my philosophical convictions.

I am grateful to the editors of *Lychnos, Centaurus, Isis*, and *Studies in History and Philosophy of Science* for permission to use, in chapters 6, 7, and 9, revised versions of articles originally published in their journals.

The staff of the following institutions allowed me access to their manuscript collections: the Oxford Geology Museum; the Berlin Geology Museum; the Geological Society of London; University College, London; the British Library; and the University Library in Geneva. Support for this project came from the National Science Foundation and a Virginia Tech Humanities Center Summer Stipend. The History and Philosophy Department of the University of Melbourne provided a hospitable environment for the preparation of the penultimate draft of this manuscript. The problems of doing research in southwest Virginia were alleviated by the willingness of Lowell Thomas and John Hayes to make repeated trips to the interlibrary loan office, and Becky Cox willingly entered a series of revisions on the word processor.

I

The Conceptual Foundations
of Geology

In this book I attempt to answer questions that puzzled me as an undergraduate studying geology in the 1960s, when foundational issues were crowded out by the mass of material to be covered. What were the conceptual foundations of geology? When and how had they been established? What aims should geologists pursue? What methods should they use to achieve those aims? What were the most appropriate entities for constructing theories about the earth? In Lyell's words, I want to reconstruct the "principles of reasoning" that geologists have used (Lyell to Murchison, in Lyell 1881, 1:234). I have chosen to tackle these questions historically, in part because of personal preference, in part because of a conviction that scientists are particularly explicit about foundational issues in periods of rapid disciplinary change. Subsequently, presumptions about a discipline's aims and methods become internalized and hence more difficult to detect.

I have accepted the conventional wisdom that the period in which the conceptual foundations of geology were laid down is the half century between 1780 and 1830.[1] Since that time, of course, geology has made immense strides forward and, particularly with the advent of plate tectonics, has acquired a theory of a scope and precision hitherto only dreamt of. But many of the basic issues that were debated in the classic period are still debated today, and many of the agreements that were reached about the aims and methods of the discipline (though not about its specific theories) continue to be accepted.

THE AIMS OF GEOLOGY: CAUSAL AND HISTORICAL

Unlike many who have inquired about the aims of geology, I have come to believe that geologists do not all share a single univocal aim. While they do share an interest in problems posed by the earth's regularities, the answers they seek are of two very different kinds. One is historical: geology should describe the development of the earth from its earliest beginnings to its present form. The other is causal: geology should lay out the causes operating to shape the earth and to produce its distinctive objects. This distinction corresponds closely to the distinction between "historical geology" and "physical geology."

Many argue that the first answer is preeminent, that the geologist aims only to reconstruct the past. Others accept that the geologist pursues more than one aim but argue that, when all is said and done, the other aims are subordinate to the task of historical reconstruction. Historical geology, on this view, integrates all the other subdisciplines of geology. Geologists study crystallography, mineralogy, petrology, sedimentology, structural geology, paleontology, geomorphology, and geophysics in order to gather the data and generalizations necessary for the reconstruction of the historical record. On the basis of an examination of geological literature and conversations with geologists, David Kitts (1977, 5) concluded

> that even among those geologists who are mainly concerned with the attempt to originate new generalizations, these generalizations are regarded as means to an end rather than as ends in themselves, the end being the construction of a chronicle of specific events occurring at specific times. With all the emphasis in recent years upon "earth science" and the theoretical physical-chemical foundation of geology, the primary concern with specific events still dominates our discipline.

In sum, geology is "radically historical" (Kitts 1977, xvii). Many, perhaps most, geologists agree. Anthony Hallam (1978, 3), himself committed to the importance of causal theories in geology, nonetheless cites Kitts and echoes G. G. Simpson's terminology when he states that geology is a "historical, non-experimental science concerned with past configurations of the earth. It deals with successions of unique, strictly unrepeatable events through time." Even philosophers, in their rare passing references to geology, adopt the same position. Ernst Nagel (1961, 550), for example, asserts that "a geologist seeks to ascertain . . . the sequential order of geologic formations, and he is able to do so in part by applying various physical laws to his materials of study." Assuming that the laws used by geologists must be physical laws, Nagel continues, "It is not part of the geologist's task, *qua* geologist, to establish the laws of mechanics or of radioactive disintegration which he employs in his investigations."

All this notwithstanding, many geologists do aim to understand the causal processes that shape the face of the earth; they see "process geology," as they term it, as central. To an extent often unappreciated by the casual observer, geologists do more than reconstruct the earth's past. They seek to identify and classify the minerals and rocks that make up the earth's crust; to analyze how sedimentary rocks are deposited, how igneous rocks are formed, and how both can be transformed into metamorphic rocks; to understand how mountains arise and are destroyed and why the earth is divided into oceans and continents; to identify and classify the fossil remains of extinct beings; and to grasp the patterns of their speciation and extinction. All these activities contribute to the formulation of *geological* laws—geological because they are couched in terms of geological entities, like basalts and trachytes, island arcs and geosynclines, plates and subduction zones. In the wake of the success of plate tectonics, some geologists even began to insist that causal theories were more central to geology than historical reconstruction. Indeed, Kitts and Hallam were responding to vocal demands that geology departments should revise their syllabuses to give a larger role to causal theories.

The distinction between historical and causal theories is not new, but goes back to the origins of the discipline, as we shall see in the body of the text. Even in the heyday of historical geology, the mid-twentieth century, some geologists argued the importance of causal theories. In the early 1950s, J. Tuzo Wilson (1954, 138–39) contrasted the two concepts of geology. The study of the earth, he said, was carried out by two distinct groups, separated by a "great barrier." Not only did each group look at different features of the earth, using very different techniques, but "the whole philosophy of their approach was different." In a tone that suggested where his sympathies ultimately lay, Wilson praised the one group "for their assiduity, care, and often courage in collecting a monumental volume of data about those detailed surface features which are exposed for them to see." But he regretted that by the very nature of their disciplinary orientation they were constrained to be "more concerned with surface effects than with the nature and causes of the deep and fundamental processes." The other group, by contrast, struggled to understand these deep and fundamental processes. But they had failed because they were "unable to analyze or make use of the field data." Ironically, this latter group agreed more about "the nature of the nuclear processes which heat the sun and characterize the atoms of different elements" than about "those slow movements within the earth which have generated its powerful magnetic field, developed its mountains, and given structure to its crust.''

Let me hasten to say that I do not draw attention to this distinction because I harbor ulterior motives of reforming the discipline. I do not wish to argue that historical geology is more important than causal, or vice versa. I have no desire to legislate how any individual contemporary geologist (or for that matter the entire geological community) should construe the aims of their enterprise. Nor do I intend to criticize geologists for allowing the coexistence of two aims, for such coexistence strikes me as probably the rule (and a good one) rather than the exception in most scientific disciplines.

My purpose, rather, is to clarify what is involved in the two conceptions of the subject, to describe how they emerged historically, and to take up the question of the extent to which they are independent or mutually supportive. Let me begin by making it clear that the distinction is *conceptual*. Criticisms leveled at it on sociological grounds miss the mark. Thus the fact (if fact it be) that *most* geologists believe that geology is primarily historical does not tell against the point that *some* have articulated a different aim and that this needs to be reckoned with. Nor does the point that many geologists pursue both aims, though perhaps not simultaneously, mean that there is no difference between the two aims.

How does the science produced by a geologist aiming to produce causal theories differ from that produced by a geologist aiming to reconstruct the past?

A scientist who seeks to develop a causal theory aims to establish general laws connecting causes and effects. Several prominent geologists have denied that such laws exist in geology, claiming that the laws geologists use are drawn from physics, chemistry, or perhaps biology (Leopold and Langbein 1963; Simpson 1963). This is either false or confused because there are plenty of geological laws (Watson 1969). An eighteenth-century instance might be "melts cool to form glassy minerals"; a nineteenth-century instance, "basalt flows under the ocean create elevation craters"; and a mid-twentieth-century one, "up-welling molten materials at the mid-ocean ridges form rigid plates." These laws, as I said, are *geological* because the entities and the processes are geological (at least in the latter two examples, once geology had become clearly differentiated from other disciplines). Of course, in the fullness of time these laws may be reduced to those of physics and chemistry, but that is true of any science. And, of course, geologists also employ the laws of physics and chemistry, but that, too, is true of any science.

Note that causal theories have a temporal dimension, and in some cases refer to events that took place in the past. The paleontologist who investigates patterns of extinction and speciation in order to construct a general evolutionary theory is perforce dealing with a process that takes place in history. This is sometimes taken to show that *all* geological theorizing is historical. And so in a certain sense it is. But this is not the very strong sense of historical that writers such as Kitts have identified, for the particular time at which the events took place is immaterial to the theorist. Claiming that geology is historical because geologists develop theories that have a temporal element misses the point. On such an interpretation, all sciences are historical.

For the historical geologist, the time sequence is paramount. As usually construed, the historical geologist aims to reconstruct a sequence of unique events. This remains a theoretical enterprise, for, as Kitts (1977) has shown, the reconstruction of such a sequence is ineliminably theoretical. But the end point is a series of particular claims, a chronicle in short, not a general theory. For Kitts and others, the contrast between causal and historical is very sharply drawn.

Indeed, it might be said that it is too sharply drawn. No room is left for a middle way, the formulation of laws of historical development. Oldroyd has contrasted the reconstruction of past events and the formulation of historical laws, arguing that the latter is not a historical enterprise in the sense intended by geologists and historians. "It is the interest in the unique historical events," Oldroyd (1979, 193) says, "rather than general historical laws, that is the hallmark of the historian. . . . Let us, therefore, distinguish between historical explanations, and those that require knowledge of a set of antecedent circumstances plus certain laws of change or development. . . . Let us refer to [the latter mode of explanation] as the *genetic* mode of explanation, emphasizing that it is substantially different from that which is characteristically employed by the historian [and the modern historical geologist]." "Genetic" theories (theories of historical development) are just a subset of causal theories. Traditionally they played an important role in geology, and, as we shall see in chapters 2 and 3, constituted the major aim of most seventeenth- and eighteenth-century cosmogonists. Indeed, some prominent geologists continue to advocate them, an advocacy that seems to me well-founded (Bowen 1928). But I agree with Oldroyd that genetic theories are not what most geologists have in mind when they say that their aim is to reconstruct the history of the earth. For the rest of the introduction, I shall restrict my discussion to historical and causal geology.

There is another way of getting at the difference between causal and historical, and that is to look at the entities that geologists treat as central in the two instances. In causal theories, the crucial entities are "natural kinds"—that is, classes of objects with some nontrivial property or set of properties in common (their essences). They are distinct, timeless, and immutable (Hull 1976, 1978). As an example, consider the geological kind "basalt." Every instance of basalt will have a common property—namely, a common chemomineralogical composition. It is this that allows us to call the substance "basalt" and to identify specimens when we find them. It is this that allows geologists to construct general laws about the kinds of igneous activity in which basalt features, whether they be nineteenth-century theories about the connection between basalt and elevation craters or twentieth-century theories about the connection between basalt and a characteristic "Hawaiian-type" of volcanic eruption. Other geological kinds would be feldspars, geosynclines, plates, and batholiths.

In historical reconstructions, the crucial entities are unique historical events or, more precisely, the rocks that were formed during a particular time period.[2] At the end of the eighteenth century, Werner defined a "formation" as the collection of rocks formed during a certain period. For him, the formation was the most crucial geological entity. The definition, with minor variations, is still standard; so, too, is the belief in the centrality of the concept to historical geology. Formations are the basic unit of stratigraphy, intermediate in scale between the stratum and the system. Each formation is unique, limited to a particular period of time, quite unlike a natural kind. It cannot function in scientific laws. The historical geologist aims to reconstruct the series of formations, of unique events.

One more feature of historical geology should be noted. Geologists rarely assert causal connections between one formation and the next. At least within the period I am discussing, geologists do not assert a causal connection between the deposition of the rocks of the Silurian and the Devonian or between those of the Jurassic and the Cretaceous. In this sense, historical geology differs from history proper. In history, sequences of events are determined with an eye to asserting implicit causal connections. The historian might claim, for example, Stuart sympathies to Catholicism caused a Puritan counterreaction. At least in the early nineteenth century, the historical geologist spent little energy wondering about the causal connections between the events that he reconstructed.

Now there might be a nice symbiosis between the two different aims for geology. If this were so, the distinction between them would be conceptually important, but it would do little to shape the discipline. Thus, if geologists constructed causal theories by generalizing from the chronicle of past unique events and if, conversely, they relied on contemporary causal theories to reconstruct past events, then the two aims would be mutually reinforcing. I shall argue, however, that the aims do not necessarily, or even usually, reinforce each other. In the course of the text, we shall see that nineteenth-century geologists reconstructed the earth's past without using the full repertoire of geological causes, and that causal theories were not based on historical reconstruction. This will become clearer in the next section, but perhaps drawing a parallel between geology and biology will help. Historians of science agree that Darwin did not arrive at his theory of evolution by contemplating past forms of life. On the contrary, he had to explain away the paleontological record as being, on first sight, at odds with his theory. The empirical warrant for evolution came from contemporary evidence—biogeography, for example—not from the earth's past. Conversely, the success of Darwin's theory did not cause a revolution in paleontology, which continued on its traditional way. Thus the historical and causal theories of life on earth developed largely independently. Similarly, human history and the social sciences are pursued largely independently, in spite of periodic attempts to change this state of affairs.

Finally, we find that, to a first degree of approximation, causal geologists are more sensitive to and more willing to exploit theories developed in other domains than are historical geologists. Their emphasis on causes makes them sensitive to the problems of consistency between their theories and those of other scientists. Historical geologists, by contrast, although they need theories in order to reconstruct past events, often find in practice that the theories they use are rather low-level and hence not as sensitive to developments in the other sciences.

THE METHODS OF GEOLOGY

So much for the aims of geologists. But how were they to achieve these aims? What methods ought they to select? Did the methods differ for causal and historical geologists? How should they relate their evidence and their theories?

As geologists in the eighteenth and nineteenth centuries construed it, this amounted to asking how to relate causes and effects. If the aim was to produce causal theories for geology, then the problem was how to determine the causes that shaped the earth's surface. If the aim was to reconstruct the sequence of events making up the past of the earth, then the problem was to determine the past effects that were the record of these events.

There was no single geological method. Geologists, like other scientists, had a whole spectrum of methods from which to choose. Before describing these methods—a necessarily lengthy description, for here lies the crux of the "principles of reasoning" in geology— I want to lay to rest a problem that has bedeviled many discussions of the role of theory and evidence—namely, too sharp a contrast between "speculators" and "empiricists." All the geologists in the eighteenth and nineteenth centuries expected theory or "system" to play a role in geology. And they all demanded that theory should be warranted by evidence. The point at issue was not whether to opt for theory or for fact gathering. Rather, it was the relationship between theory and facts. By the late seventeenth century, few would have disagreed with the claim of cosmogonist John Woodward (1695, 1), who said, "From a long train of Experience, the World is at length convinc'd, that observations are the only sure Grounds whereon to build a lasting and substantial Philosophy." From then on, the question for geologists was how to relate the observations they gathered to the "philosophy" they constructed.

But many historians of science hold that scientists' methodological claims are mere window dressing, *machines de guerre* employed to achieve other ends, playing no substantive role in the attitudes they adopted to theories. This seems mistaken. We need to guard against the tendency to project our own attitudes to methodology back on our predecessors. Contemporary education in the natural sciences places little emphasis on methodology construed as the investigation of the relation of theory and evidence (although, of course, considerable emphasis on methodology construed as specific techniques and procedures). And since scientific education has become very specialized, few encounter the issue in other contexts—say, in philosophy courses. Nor does the highly formalized style of scientific papers allow for methodological discussion. It was quite otherwise in the seventeenth, eighteenth, and nineteenth centuries. Any scientist was familiar with classical writings and discursive scientific texts. These frequently contained lengthy sections on methodology, which shaped the way scientists thought about their work.

I shall begin with the methods of the causal theorist. Not only have these been much more thoroughly analyzed than those of the historical geologist, but they were worked out prior to the development of methods of reconstructing history. These methodologies were not peculiar to geology. Natural philosophers, natural historians, and chemists all addressed methodological questions. Over time, as natural philosophers became clearer about the nature, strengths, and weaknesses of the various alternatives, the popularity of the different methods waxed and waned, changes that were reflected in geology (Laudan 1981; see also Blake, Ducasse, and Madden 1960; Losee 1972).

Geologists also realized that they, like other scientists, had some evidential problems specific to their own field of study. As these became clearer, geologists began to judge methodologies by their adaptability to these special problems, as well as by their general standing among scientists. Lyell, in particular, proved quite ingenious at modifying standard methodologies to fit geology. In chapter 9, I shall argue that uniformitarianism was, and was recognized to be, an attempt to modify a standard method—the method of *vera causa* or true causes—to the particular subject matter that exercised geologists.

Among the special problems for geologists were the inaccessibility of many geological processes, the relevance of experimental evidence, the long time period required for many geological changes, and the complexity of geological phenomena. By the early eighteenth century, most geologists recognized that many geological phenomena resulted from submarine and subterranean changes. Since these were plainly inaccessible to any human observer, geologists had to decide what could be legitimately claimed about them. Throughout the eighteenth century, mineralogists and cosmogonists assumed that analogies between laboratory experiments and geological processes were unproblematic (see chapters 3 and 5). But from the late eighteenth century on, some geologists began to argue that natural and laboratory conditions were so disparate that using these analogies led to serious misconceptions about the nature of processes taking place in the earth. They compared the geologist to the astronomer, who had largely to wait passively to observe eclipses or meteor showers without any possibility of reproducing, much less actively interfering with, such events.

For much of the eighteenth century, geologists assumed that the earth had a short past and that contemporary causes were efficacious. Hence the time scale on which geological events took place was no impediment to theorizing. But the success of historical geology meant that, although geologists remained reluctant to put a specific age on the earth, most of them did conclude, or at least entertain the possibility, that many events took place over very long periods of time. Here the analogy with astronomy broke down. For, unlike eclipses and meteor showers, many geological processes took far too long for their completion to be observable. Of course, the astronomer also studied some long-term processes, such as the return of comets. But the ratio of long-term to short-term events was much greater in geology than in astronomy—or at least this was what many geologists believed. Those geologists who reckoned past transformations of the globe were much faster and more furious than current ones were of one mind with those who admitted no such slowing down in terrestrial nature: the few thousand years of recorded human observation were quite insufficient for observing from start to finish many geological changes—the deposition of large marine sedimentary formations, for example.

Finally, the enormously diverse changes wrought beneath and on the earth's surface were the work not of a few causes whose separate effects could be distinguished easily but of many causes that were often in highly complex interaction with one another. The transformation of a region from ocean into continental land might well be considered as a single event, but it could only be explained as an overwhelmingly complicated effect produced by an intricate sequence of operations conditioned at every stage by the cooperation of a whole battery of natural agencies.

Geologists, like other scientists, considered a spectrum of five widely discussed methods for determining the relations between causes and effects—the methods of hypothesis, analogy, enumerative induction, eliminative induction, and the *vera causa* method. At one extreme of the spectrum was the method of hypothesis. Full-fledged hypothetico-deductivism was very popular in the seventeenth century, particular among the corpuscular philosophers (Laudan 1971, chap. 4). Cosmogonists, many of whom accepted the corpuscular philosophy, used it to construct genetic accounts of the earth's development. The method slipped from respectability in the eighteenth century, though a minority of scientists continued to advocate it, including the Swiss physicist George-Louis Le Sage (1724–1803) and the English psychologist David Hartley (1705–

1757). Within geology, James Hutton (1726–1797) relied on it time
and again, a matter of some embarrassment to his friend John
Playfair (1748–1819), who played down its importance in his *Illustrations of the Huttonian Theory of the Earth* (1802) (see chapter 6).
In the nineteenth century, Pierre Prevost (1751–1839) in France and
Thomas Young (1773–1829) in England took up the cudgels on
behalf of the method of hypothesis (Olson 1975; L. Laudan 1981).
Those who subscribed to this method judged any hypothesis that
was compatible with the evidence to be at least probable. The great
advantage of this method, of course, was that it warranted inferences on the strength of a small and accessible range of data.

However, the trouble with the method of hypothesis, as virtually
everyone realized, was its highly inconclusive character. The fact
that a hypothesis was sufficient to explain a limited body of evidence
did not rule out the possibility that the hypothesis was very wide of
the mark. Because many contrary hypotheses might be compatible
with the same body of evidence, mere compatibility between a
hypothesis and a limited number of observed effects was thought to
be no reliable guide to the truth of the hypothesis. Of all the
available methodologies, the method of hypothesis was the least
reliable. Lyell, who had severe reservations about it on this score,
saved some of his most splenetic language for those among his
predecessors whom he believed to be devotees of the method.
Talking about John Woodward, Lyell (1830–33, 1:39), said, "Like
all who introduced purely hypothetical causes to account for natural
phenomena, [he] retarded the progress of truth, diverting men from
the investigation of the laws of sublunary nature, and inducing them
to waste time in speculations." Therefore, Lyell urged that some
check must "be given to the utmost license of conjecture" (1:186) in
speculating on the causes of geological phenomena.

At the other extreme were the eliminative inductivists, who
maintained that the data could be used to prove the truth of a
hypothesis, by refuting *all* its conceivable rivals. This tradition,
often associated with Francis Bacon and his conception of crucial
experiments, was an influential one. The putative certitude that it
offered had great appeal. James Hutton, for example, used it to
warrant his theory that rocks had been consolidated by heat. He
dismissed what seemed to be the only possible rival theory—the
theory that they had been consolidated by water—and then without
further ado announced that his theory had won the day (see chapter
6). At a later period, Charles Darwin thought he could use the
method to good effect. Attempting to formulate a theory of eleva-

tion, with special reference to the parallel "roads" or shelves running along the sides of the Scottish valley of Glen Roy, he introduced three possible hypotheses. Having dismissed two of them, he claimed—admittedly uneasily—the third (his own) as the only acceptable option: "Having now discussed those views which cannot be admitted,—a method of reasoning always most unsatisfactory, but necessary in this instance from the high authority of those who have advanced them,—I will consider some other appearances, which will perhaps throw light on the origin of the shelves" (Darwin 1839, reprinted in Barrett 1977, 99).

But the apparent certainty endowed by the method frequently turned out to be spurious. The great problem, of course, was to produce a truly exhaustive list of all the possible rival hypotheses. Since this difficulty was widely appreciated, many scientists regarded eliminative induction with suspicion. The Irish chemist Richard Kirwan rebuked Hutton for employing this method without giving any positive arguments for his theory (see chapter 6). Darwin himself was embarrassed when Agassiz proposed his glacier-lake hypothesis. This exposed the fallacies in Darwin's reasoning, showing that Darwin had not exhausted all the causes capable of producing the "roads." Darwin remarked, "My error has been a good lesson to me never to trust in science to the principle of exclusion" (quoted in Barlow 1958, 84; see also Hull 1972, 1973; Rudwick 1974a).

The other widely discussed inductive methodology was that of enumerative induction. On this view, the only legitimate theories are those that emerge as simple generalizations from the data. Specification of hypotheses to guide the collection of data or to be tested by the data was disallowed. Not only was this method popular in the general scientific community, particularly in the eighteenth and early nineteenth centuries, but it was the preferred method of many geologists. G. B. Greenough effectively succeeded in making enumerative induction the official methodology of the Geological Society of London (Rudwick 1963; Rachel Laudan 1977b) (see chapter 7).

However respectable, there was nonetheless a major problem with enumerative induction, particularly as far as geology was concerned. As many critics of the method realized, it was much too restrictive. If the aim of geologists is to discover the causes leading to the geological effects that they see around them, enumerative induction is of little help. For the method to work, the scientist must be able to observe relevant conjunctions of cause and effect in a given situation, then generalize these conjunctions. Such conjunctions are very hard to observe where geological causes and effects are concerned. In many cases, the causes that produced the effects that we now observe acted in the past; they are no longer directly observable. In other cases, causes that can be observed at present are acting so slowly that their effects cannot yet be detected. Other processes are taking place in the interior of the earth and are equally removed from observation. Strict adherence to the method of enumerative induction meant that most of the phenomena that geologists wished to examine would not be amenable to scientific investigation. Hence Lyell (1827, 440–41), who fully appreciated these problems, encouraged his readers to "entirely disavow the influence of that fashion, now too prevalent in this country, of discountenancing almost all geological speculation."

That left the methods of analogy and true causes. Many geologists thought that these avoided the methodological pitfalls of the extreme ends of the spectrum. They regarded the method of analogy as intermediate between the methods of hypothesis and induction. Whereas enumerative induction only licensed inferences from some known event to other events that were similar to it in relevant respects, the method of analogy warranted inferences from known events to unknown ones that were merely similar or analogous to the known ones. If one cause-effect chain was "known," and a similar effect (or cause) also known, then from these three "knowns," the method of analogy—operating like the mathematical method of ratios—provided information about the fourth, unknown variable. In the eighteenth century, cosmogonists used the method to warrant cosmogonies based on laboratory experiments (see chapter 3). In the nineteenth century, geologists employed it to reconstruct past environmental conditions (see chapter 7). Lyell frequently resorted to this method, particularly when discussing why reliance on evidence from paleontology was warranted. Of course, the method was not without its difficulties, not the least of which was the problem of deciding how "similar" objects or events had to be before the scientist was justified in drawing extensive analogies between them.

The final method was the method of "true causes," or the *vera causa* method. The source for this was none other than Sir Isaac Newton. Newton had not only produced spectacularly successful scientific results but had offered hints in his famous "Rules of Reasoning" in the *Principia* about how others could do likewise. The first of these rules—"We are to admit no more causes of natural things than such as are both true and sufficient to explain their appearances"—became known as the "principle of true causes," or the *vera causa* principle. Historians of science have long debated the extent to which Newton followd his own rules, concluding that in general he honored them more in the breach than in his day-to-day practice. But there is no doubt that they were taken quite seriously in the eighteenth and nineteenth centuries (Butts and Davis 1970; Laudan 1981).

The Scottish philosopher Thomas Reid (1710–1796) argued particularly strongly for the utility of Newton's "Rules" for scientific research. In his *Essays on the Intellectual Powers of Man* (1785/1969, 88), Reid wrote,

> When men pretend to account for any of the operations of nature, the causes assigned by them ought, as Sir Isaac Newton taught us, to have two conditions, otherwise they are good for nothing. First, they ought to be true, to have a real existence, and not to be barely conjectured to exist, without proof. Secondly, they ought to be sufficient to produce the effect.

Reid's first requirement—that any putative cause must be known "to have a real existence"—was interpreted as an insistence that we must have direct observational evidence that causes of the kind assumed do actually exist. Reid was not, however, requiring that the cause be observed operating in the case to be explained; rather, he was making the weaker demand that there be evidence that such causes be known to exist in some circumstances. His second requirement was that the cause be adequate to produce the purported effect.

Eighteenth- and nineteenth-century scientists of all persuasions held these strictures in high regard. Richard Kirwan (1799, 1–2), for example, asserted, "In the investigation of past facts dependent on natural causes, certain laws of reasoning should inviolably be adhered to." He went on to name three: "no effect shall be attributed to a cause whose *known* powers are inadequate to its production"; "no cause should be adduced whose existence is not proved either by actual experience or approved testimony"; and "no powers should be ascribed to an alleged cause but those that it is known by actual observation to possess in appropriated [*sic*] circumstances." Having used the method of hypothesis to construct a cosmogony in which a flood played a major role, Kirwan (1799, 6.) elevated the inference to *vera causa* respectability by invoking Moses' eyewitness testimony that a Deluge had actually occurred.

Playfair, in his treatment of the Huttonian theory, went to some pains to point out that the causes that Hutton was citing were *verae causae*. He argued that it would be wrong to accept Hutton's theory of subterranean heat if the only evidence for it were "the geological facts which it is intended to explain" (Playfair 1802, 91). It was legitimate because there was *independent* evidence for the heat's existence in "phenomena within the circle of ordinary experience; namely, those of hot-springs, volcanoes, and earthquakes." Lyell, too, thought that the *vera causa* method, which offered a nice middle way between too much and too little theorizing, was superior to the available rivals. He suggested how the *vera causa* principle might be modified to address the special evidential problems of geology and thus formulated his famous principle of uniformity (see chapter 9).

Historical theorists used the same range of methods, but inverted them, for they aimed to reconstruct effects rather than causal processes. This difference in emphasis is reflected in the different kinds of evidential warrant to which the two kinds of geologist appealed. Causal theorists (including genetic theorists) wanted, at the end of their investigations, to have determined the causal processes that shaped the face of the earth. These were usually processes of erosion, consolidation of loose material into rocks and minerals, and changes of level of land and sea so that new rock appeared above the level of the ocean. They were quite willing to turn to extrapolations of physical and chemical theories, to laboratory analogies as well as to field observations, in order to provide the evidential warrant for their causal claims, but those causal claims had to be couched in terms appropriate to geology. Historical geologists wanted, at the end of their investigations, to have determined the series of unique events that formed the earth's history. Since these were primarily recorded in the rocks (and to a lesser extent by human testimony), it was to the field that they turned for their evidential warrant. They were willing to use causal principles derived from any science to help them interpret the rocks. Whether or not the theories were geological mattered little, so long as they could reconstruct a sequence of unique events.

Both historical and causal geologists were much concerned with the methodology of correctly identifying natural kinds or, in other words, with the theory of taxonomy (see chapters 4 and 7). Until recent years, this has been largely neglected by philosophers, who, dealing with the sparse ontologies of physics, have been able to leave the question to one side. Biologists and philosophers of biology have been forced to tackle the problem of correctly dividing up the diversity of the natural world, reviving an older tradition. During the eighteenth century, biologists had wanted assurance that they had correctly identified and organized plants and animals, a concern that led to their preoccupation with taxonomy. Mineralogists tried to adopt the biologists' methods. In chapter 4, I shall discuss this at some length, indicating why they eventually concluded that the methods that had worked so well for the plant and animal kingdoms failed so dismally when applied to the objects of geology.

Historical and causal geology moved ahead with a fair degree of independence. Although historical geologists required theory to reconstruct the succession, as well as to reconstruct past conditions, they did not need the complete range, or even a large proportion of causal geological theory. In the 1820s, for example, historical geologists paid little heed to the theories of mountain elevation that causal geologists were propounding because they were irrelevant to establishing the succession of rock formations (see chapter 7). Conversely, those causal geologists had little to learn from a history that was told in terms of a sequence of formations; since formations were not natural kinds, they did not employ them in their generalizations about causes. Thus, theorists of mountain elevation largely ignored the fine problems of working out the details of the geological record (see chapter 8).

Of course, geologists wanted consistency between their causal and historical theories, so debates about, say, the kinds and intensities of causes were common. And, presumably, they eventually wanted a fully integrated causal-historical story marked by the kind of reconciliation that at present seems to be taking place between evolutionary theory and paleontology. But in the short term in which geologists worked, it was not only possible to pursue historical and causal geology independently but they often preferred to do so.

THE THEORIES OF GEOLOGY

Thus far, I have been talking in a rather abstract way about the aims and methods of geology. It is now time to give a brief overview of the succession of geological theories resulting from the pursuit of these aims according to the methods described. I shall describe the changes between the late seventeenth century, when mineralogy and cosmogony combined to provide unified genetic theories of the earth's history, and the early nineteenth century, when, under the banner of geology, scientists pursued causal theories largely independently from strictly historical reconstructions of the earth's past.

In chapter 2, I sketch the "commonsense knowledge" of the mineralogist and show how this was integrated into cosmogonies (genetic causal theories). In chapter 3, I outline the social conditions that allowed a particular kind of cosmogony—that based on the chemical theories of J. J. Becher and G. Stahl—to flourish in eighteenth-century Germany, Sweden, and France. In chapter 4, I look at the eighteenth-century efforts to apply taxonomic methods that had worked well for botany to the mineral kingdom and show how mineralogists' realization of their inapplicability prepared the way for them to accept historical entities (formations) as well as the natural kinds (minerals and rocks) of classic taxonomic theory. In chapter 5, I examine how Abraham Gottlob Werner drew on both the Becher-Stahl cosmogonic tradition and the work in mineral taxonomy to construct a geological ("geognostic") theory that opened the way to a separation between causal and historical geology. In chapter 6, I turn aside from the main narrative to look, through the eyes of a mineralogist in the Becher-Stahl tradition, at James Hutton's theory of the earth in order to see why his theory received such a lukewarm reception. In chapters 7 and 8, I trace out the social and intellectual ramifications of the successful extension of Wernerian historical and causal geology. I argue that this Werner-ian "radiation" was the dominant tradition in the first quarter of the nineteenth century. In chapter 9, I show how Lyell tried to provide new methodological underpinnings for causal geology, necessitated by the recognition of the length of the earth's history. I conclude, in chapter 10, by contrasting my history with the "received view" of the history of geology, by reflecting on the historiography of geology, and by considering some of the implications of my story for scientific change in general.

THE AIMS OF HISTORY OF SCIENCE

The themes of this book—the development of the aims, methods, and theories of geology—put me squarely in the camp of those who believe that what makes science worthy of study, what gives it interest, are scientists' beliefs about the world. It is a truism, but one worth remembering, that what sets natural science off as a distinct field of human activity is the success with which scientists have investigated the natural world in the past three or four centuries. Some critics dismiss this success as merely apparent, as a giant confidence trick that should be unmasked. But for those historians of science who disagree—and they are the majority, I believe—the central problem is to trace the history of scientific ideas and, if possible, to shed some light on why they seem so accurately to describe the world.

But I am well aware that setting out to write an extended account of the cognitive development of a discipline marks me as somewhat dated and out of touch with current trends in history of science, particularly social history of science and microstudies. So some brief explanation is in order.

By saying that understanding scientists' beliefs is the first order of business, I am not thereby committing myself to a rigid internalism, nor to the thesis that human actors are unimportant in science, nor yet to the view that there are "disembodied" ideas. The social history of science must play second fiddle to the intellectual history of science because only when we have traced the development of scientists' ideas do we know what needs to be explained—namely, the causes of those ideas, the conditions that conduce to their formulation, discussion, and propagation, and the reasons for their success. But second fiddles play an important supporting role. The causes proposed to explain scientists' beliefs may be intellectual or social. That is an empirical question. Furthermore, scientists work in particular social structures that may or may not contribute to the construction, consideration, and transmission of theories, and one task for the historian is to understand that context. In chapters 3 and 5, I shall consider the social context surrounding the transition from mineralogy to geology.

As to microstudies, they can illuminate the fine structure of the generation of scientific ideas in an unparalleled way, besides offering a splendid opportunity for historical virtuosity. But historians must select microstudies according to some principles. They need to be able to say what feature of science their studies illuminate. Among other things, this presupposes that the author has a good grip on the history of theories of the science in question. For some disciplines and periods, such histories may now be readily available—physics at the time of Newton or biology at the time of Darwin, for example. But the historiography of geology is still in a relatively primitive and unexplored state. To assume that nineteenth-century British geology—which is the focus of most microstudies in the history of geology—represents a typical example of geology's aims, methods, and problems is, to say the least, open to argument. Detailed studies based on an unquestioned, but ubiquitous, assumption of this kind have to be set in their wider context with great care.

Finally, to admit that my desire to understand modern geology constrains the historical issues I address is to run the risk of being accused of "Whiggism." In spite of some recent efforts to clarify this notion, and hence render it inoffensive (Hull 1979; Hall 1983), many still assume that approaching the past with present concerns in mind falsifies the past. A presentist approach, it is insinuated (though rarely argued), so blinkers the historian that anyone who falls prey to it loses the capacity to understand the perspectives of the historical actors, judging earlier theories by the extent to which they do (or do not) approach our present beliefs.

But to conflate selection and judgment is simply wrongheaded. Selecting our historical problems in light of their bearing on contemporary problems does *not* mean that we thereby render ourselves incapable of seeing those historical problems in historical context. Quite the contrary. Historians have to explain how that context shaped the formulation of and answers to the problem. They have to understand the situations in which historical actors found themselves. Choosing those problems in the history of geology that seem to have been particularly significant in the formation of modern geology does *not* commit me to interpreting those episodes in terms of the *standards* of modern geology.

2
Mineralogy and Cosmogony in the Late Seventeenth Century

Eighteenth-century geologists, including those with such apparently diverse theories as Werner and Hutton, were of one mind about many fundamental issues. They believed that the outer portion of the earth was made up of four major mineral classes—the "earths," the "metals," the "salts," and the "bituminous substances." They believed that these classes could be distinguished by their reactions to heat and water. They believed that these minerals had once been fluid and had later been rendered solid by the removal of heat or water. Taken together, these doctrines constituted the "common sense" of geology. With hindsight, we can see them for what they were—namely, a series of low-level theoretical claims. But at the time, they were perceived as the unproblematic descriptive categories of the field. They simultaneously shaped mineralogists' perceptions of their data and demanded explanation by higher-level theories. Furthermore, since consolidation, as the transition from fluidity to solidity was called, was so central to this conception of mineralogy, consolidation constituted a major problem, if not *the* major problem, to be tackled until the end of the eighteenth century.

This consensus about the conceptual framework, if not the specific theories, of geology had been in place since at least the late seventeenth century. It crumbled only in the early nineteenth century. To appreciate the debates that occurred in the transition from mineralogy to geology and the matrix of ideas that underlay the theoretical changes between 1780 and 1830, we need to reconstruct the "common sense" of the eighteenth- and nineteenth-century mineralogist. In this chapter, I shall trace the roots of that common sense. I shall then show how two of the great rival systems of the seventeenth century—the mechanical and the chemical philosophies—attempted to incorporate this common sense. Adherents of each system constructed genetic theories of the earth based largely on the method of hypothesis. I shall conclude by looking at typical examples of each type of theory, the cosmogony proposed by Descartes within the mechanical tradition, and the cosmogony proposed by Becher within the chemical tradition.

But first, a word about the terms *mineral, mineralogy,* and *mineralogist* is in order since their meanings have changed since the eighteenth century (Kobell 1863; Adams 1934, 1938; St. Clair 1966; Oldroyd 1974d).

The mineralogist collected specimens from mines and quarries, and carried them home for examination in the study, the "cabinet," or, by the eighteenth century, the laboratory (Mornet 1911; Dance 1966; and Prescher 1980). Today, the term *mineral* refers to the chemical compounds making up the rocks of the earth's crust; the subdiscipline of mineralogy is restricted to the investigation of these chemicals, particularly the complex silicates. At that time, the term *mineral* referred to *all* the naturally occurring, nonliving, solid objects on the globe (and in some systems, fluids as well). Put colloquially, minerals referred to all the objects that we would include under "mineral" in the popular parlor game of twenty questions. Put in the language of the eighteenth century, minerals referred to all the objects in the third of the three "kingdoms of nature"—the animal, the vegetable, and the mineral. (Then, as now, the term was also used to refer to any valuable ore mass, but that was a secondary, nontechnical sense.) Thus, mineralogy was no mere subdiscipline. Rather, it comprised most of the nonhistorical parts of what is now geology—crystallography, mineralogy, petrology, and paleontology—as well as interpenetrating much of the domain of chemistry. As the eighteenth century wore on, its scope widened. At the beginning of the last quarter of the century, Werner (1774/1962) defined *mineralogy* as being made up of three major subdivisions that, taken together, closely approximate the scope of modern geology: oryctognosy (the identification and classification of minerals), mineral geography (the distribution of rocks and minerals), and geognosy (the formation and history of rocks and minerals).[1]

MINERALOGICAL "COMMON SENSE" IN THE SEVENTEENTH AND EIGHTEENTH CENTURIES

Any eighteenth-century mineralogist, if asked what constituted the solid parts of the earth, would have answered "minerals." By this he would have meant the four classes of minerals—metals, earths and stones, salts, and sulfurs—arranged in extended masses as rocks, veins, and strata. From the sixteenth to the late eighteenth century, the descriptions of the major mineral classes—and of rocks, veins, and strata—remained stable (table 1). Whether one looks at Georgius Agricola (1494–1555), Konrad Gesner (1516–1565), John Kentmann (1518–1574), Andrea Cesalpino (1519–1603), Ulisse Aldovandri (1522–1605), and Ferrante Imperator (1550–1625), in the sixteenth century, John Woodward (1665–1728) and Johann Scheuchzer (1672–1733) in the seventeenth century, or Carl Linnaeus (1707–1778), Johan Wallerius (1709–1785), John Hill (1707–1775), Axel Cronstedt (1722–1765), James Hutton (1726–1797), Jacques Valmont de Bomare (1731–1807), Jean Baptiste de Romé de l'Isle (1736–1790), and Abraham Gottlob Werner (1750–1817) in the eighteenth century, this generalization holds true.[2] The same was not true of the descriptions of mineral species within the major classes. In sharp contrast to the organic world, in which botanists and zoologists recognized stable species, perhaps even more stable than the higher taxonomic units, mineralogists were forever shuffling their organization of the minerals within classes. This was to have a significant effect on mineral taxonomy (see chapter 4).

The mineral classes played a pivotal role in theories about the earth's structure or history. On the one hand, mineralogists had to explain why the earth's crust was differentiated into these classes; on the other hand, they used the mineral classes to explain the gross features of the crust—its rocks, and ultimately its physical geography—and to reconstruct the earth's history.

Eighteenth-century mineralogists looked back to the German physician-natural historian Agricola as the founder of modern mineralogy. In his magisterial *De Natura Fossilium*, published in 1546, Agricola ignored the medieval lapidaries, with their catalogs of the size, color, weight, heraldic significance, medicinal use, and symbolic function of selected minerals. Instead he sought to modernize the mineralogies of the medieval scholars Avicenna (980?–1027) and Albertus Magnus (1200–1280), and thus ultimately the scattered discussions of mineralogy in Aristotle's *Meteorologica*.[3] Agricola identified approximately 600 "simple" (that is, chemically homogenous) minerals and grouped them into four major classes.

"Metals" were usually solid (though Agricola—parting company with the chemists—included mercury, a liquid, among the metals), and they melted when heated, returning to their original form on cooling (Agricola 1546/1955). The major metals were gold, silver, iron, copper, tin, and lead. Metals had been recognized as a distinct

Table 1. The Evolution of Mineral Systems from Agricola to the Early Nineteenth Century

AGRICOLA 1546		WOODWARD 1704		LINNAEUS 1735			HENCKEL ca. 1730–1744	
Earths		Earths		Stones	Calcareous		Earths	Refractory
								Fusible
Stones		Stones			Argillaceous		Stones	Calcareous
								Argillaceous
					Vitrifiable			Calcareous-Siliceous
								Siliceous
Congealed Juices	Salts	Salts		Minerals	Salts	Salts		
	Sulfurs	Bitumens			Sulfurs	Sulfurs		
Metals		Metals			Metals			
		Metallic Substances						

Table 1 (Continued)

POTT 1746		WALLERIUS 1747		CRONSTEDT 1758		BERGMAN 1783	
Earths + Stones	Gypsous	Earths	Mineral	Earths	Calcareous	Earths	Calcareous
			Argillaceous		Argillaceous		
	Alkaline + Calcareous		Dry		Siliceous		Magnesia
					Garnet kind		
			Sandy				
	Argillaceous	Stones	Calcareous		Micaceous		Argillaceous
					"Fluores"		
			Argillaceous		"Asbestos kind"		Siliceous
	Siliceous		Vitrifiable		Manganese kind		
			Rocks		Zeolites		Heavy
		Minerals	Salts	Salts		Salts	
			Sulfurs	Inflammables		Inflammables	
			Metals				
			Semimetals	Metals		Metals	

Table 1. (Continued)

WERNER 1789		PHILLIPS 1816		WERNER 1817	
Earths + Stones	Calcareous	Earths	Lime	Earths	Lime
			Magnesia		Talc
	Talky		Alumine		Clay
			Silex		Flint
	Clayey		Zircon		Zircon
			Glucine		Diamond
	Siliceous		Yttria		Barite
			Barytes		Strontian
	Heavy		Strontian		Halite
Salts		Alkalis & Alkaline Earths		Salts	
Inflammables		Combustibles		Combustibles	
Metals		Metals		Metals	

Compiled from Agricola [1546] 1955; Woodward, in Harris 1704; Bergman 1783; Werner 1789a; Phillips 1816; Werner 1817; Oldroyd 1974b (courtesy the editor of *Isis*). For purposes of simplicity, I have ignored the wider classifications of the kingdoms of nature in which mineral classifications were often imbedded. I have also ignored the finer divisions of the classifications, which often ran to hundreds of items.

The major classes of earths and stones, salts, sulfurs (or bitumens or inflammables), and metals persist throughout the period, forming the "common sense" of mineralogy. Distinguished on the basis of their reactions to fire and water, these classes shaped mineralogists' accounts of the causal processes at work. In the eighteenth century, mineralogists differentiated the earths (the major components of the rocks of the earth's crust) with increasing precision, thus imposing stronger empirical constraints on theories of the earth. (The subdivisions of the other classes, which were of less moment for geology, are not shown.)

class since antiquity. Aristotle (*Meteorologica* III, 6) had described "metals" (or "things mined," as he called them) in essentially the same terms as Agricola. Until late in the eighteenth century, many mineralogists included metal ores, as well as metals proper, in this class. The economic value of metals ensured that they, and the veins in which they typically occurred, dominated the attention of mineralogists.

The "earths" and the "stones" made up the bulk of the earth's surface; that is, they formed the extended masses known as "rocks" and were the matrix in which the other minerals were usually found. They also occurred as small individual specimens, such as geodes and gems. Mineralogists were unsure whether the earths and stones should be put in two separate classes or treated as different manifestations of the same class. The confusion persisted until the latter half of the eighteenth century. (For simplicity's sake, I shall simply talk about earths, since mineralogists generally agreed that earths and stones could only be distinguished by the relative degrees of induration, and thus that no sharp line existed between them.) Since the earths had little economic value and resisted experimental investigation, mineralogists largely overlooked them until the mid-eighteenth century. Thereafter, distinguishing different earths became a major preoccupation (see chapter 3).

Agricola followed Avicenna in describing "earths" and "stones" as proper classes, rather than as mere matrices for economically valuable minerals as Aristotle had done. Our everyday use of the terms captures Agricola's sense fairly closely. Earths could be worked with the hands when moistened, turned to mud with the addition of further water, and returned to their original state when dry. Fire affected earths little, if at all. Typical earths were the soils found in the fields. Stones were hard and dry, whereas earths were powdery. But like earths, stones were usually inert. They softened very slowly, if at all, in water, and usually resisted fire, although they sometimes melted or crumbled. Typical stones were marbles, limestones, geodes, and gems.

The two remaining classes were the "salts" and the "sulfurs" (sometimes called "bituminous substances"), which Agricola lumped together as "congealed Juices." Salts and sulfurs tended to be dry and rather hard. When treated with water, however, they dissolved or at least softened. Salts were rather resistant to fire, whereas sulfurs burned. Salts, such as rock salt and alum, roughly corresponded to our modern meaning of the terms, although they also included acids and alkalis. Sulfurs were substances such as coal and bitumen. Both salts and sulfurs occurred as isolated masses in the earth, occasionally forming extended rocks.

Note that Agricola, like his predecessors and his successors, distinguished the mineral classes by their differing reactions to heat and water. Since heat and water are the major agents of geological change, this established a continuum between the investigation of minerals in the cabinet or laboratory and the understanding of the processes of change on the face of the earth. Mineralogists quickly applied refinements and discoveries in the former to new theories about the latter. But before turning to this grander scale, we need to consider a little more of the descriptive vocabulary of the mineralogist.

Mineralogists found that some objects in the earth, by virtue of their very distinct shapes or forms, seemed to demand classification by shape rather than by their material constitution. The most important of these objects were "crystals," "figured stones" (fossils in modern terms), and "strata" (parallel layers of extended mineral masses). Mineralogists' orthogonal classification of minerals by shape, rather than substance, forms a complicating factor in the history of mineralogy and geology. Crystals were usually earths or salts. Figured stones might be composed of any of the mineral classes, but were usually earths. And strata could be earths, bituminous substances (coal), or salts (rock salt). The regular geometric shape of crystals and the resemblance between organic fossils and living beings seemed to presuppose some inward form. For millenia, geometric figures and living species were regarded as the prototypes of natural kinds. The mineralogist had to decide whether the imperfect geometric shapes found in naturally occurring crystals and the imperfect resemblances to living beings found in fossils were also natural kinds and, if so, how they were formed. Since the shapes of these objects were not explicable in chemical terms alone, mineralogists frequently invoked some sort of Platonic form or idea—usually called an "architectonic" principle or *vis formativa*—to explain the appearance of crystals and fossils in particular.

Finally, mineralogists employed a number of terms to describe large-scale mineral masses, terms that once again cut across the other divisions we have examined. "Rocks" were extended masses of minerals, usually earths or stones, but sometimes salts or bituminous minerals.[4] "Strata" were a subclass of the whole group of extended mineral masses, the rocks, their distinguishing feature being that they were layered. However, not all rocks were strata, for not all rocks were layered. Finally, "veins" were any relatively thin mineral bodies that cut across the grain of the rocks. Usually metals, veins could also be formed of earths (igneous intrusions), salts, or bituminous substances. As the site of most economically important minerals, veins aroused great interest among mineralogists.

MINERALS AND BASIC ONTOLOGIES

Mineralogists sought to explain the behavior of the mineral classes in terms of more basic ontologies. The main choices were the organic, the chemical, and the mechanical. The organic theory, which postulated that minerals grew in the earth just as plants and animals grew on its surface, was present in antiquity, flourished with the rise of Neoplatonism and Hermeticism in the Renaissance, and persisted into the eighteenth century (Oldroyd 1974e; Debus 1977). Jean-Baptiste Robinet (1735–1820), for example, argued for it strenuously in *De la Nature* (1761–66). However, I shall pay the organic tradition little attention since modern mineralogy owed more to the mechanical and chemical traditions. The mechanical theory attempted to explain the earth's mineralogy in terms of the elementary particles of the mechanical philosophy (Oldroyd 1974c). It became particularly prominent in the mid-seventeenth century, as a result of the advocacy of Descartes, Boyle, and others. Even so, it remained a minority strand in the development of mineralogy and geology, which was dominated by the chemical theory of elements or principles (Multhauf 1958). It is to that that I shall turn first.

Chemical Theories of Mineralogy

Until the Chemical Revolution of the late eighteenth century, chemists explained the natural world, including minerals, by reference to a small number of "elements" or "principles." According to Aristotle's four-element theory (*De Generatione et Corruptione*, book 2, chaps. 1–8), from which most later chemical systems stemmed, the world was composed of the elements "earth," "air," "fire," and "water." Each element was characterized by a pair of the four basic qualities—hot, dry, moist, and cold—imposed on a hypothetical substrate, the *hyle*. Thus, earth was cold and dry, water cold and wet, air hot and wet, and fire hot and dry. Changes in the qualities brought about the transmutation of the elements. All naturally occurring substances, even those that bore the name of an element, were composed of all the elements. This was true of minerals, though Aristotle though that their predominant constituents were earth and water (*Meteorologica*, book 4, chap. 8).

Another part of Aristotle's system—his theory of "exhalations"—played such an important role in chemical and mineralogical theory that it needs to be mentioned here. Aristotle postulated that the earth gave off "exhalations" when subjected to the heat of the sun (*Meteorologica*, book 3, chap. 6). The exhalations were of two kinds, the moist and the dry. Moist exhalations, if they escaped from the earth, caused rain, dew, hail, and snow; the dry, in similar circumstances, caused shooting stars, wind, thunder, lightning, and earthquakes (Eichholz 1949). But if the moist and dry exhalations were trapped within the earth, they formed minerals. "Metals" were the congealed form of the moist exhalation. "Fossiles" (the term classicists have invented to translate "things that were dug up" that is, the economically important earths) were produced by the entrapment of dry exhalations.[5] Since the moist exhalation was predominantly composed of the element "water," it followed that metals were primarily "water." "Fossiles" were predominantly "earth." Things were not quite this simple, of course. The exhalations often commingled, so that small amounts of the dry exhalation appeared in metals and vice versa.

The history of chemistry between antiquity and the seventeenth century is complex and confusing. The medieval Arab chemists suggested that Aristotle's moist and dry exhalations were mercurial and sulfurial, respectively. Paracelsus (1493–1541) revised Arab chemical theory and substituted salt, sulfur, and mercury for Aristotle's earth, air, fire, and water. He identified salt as the principle of solidity, sulfur (the equivalent of Aristotle's dry exhalation) as the principle of flammability, and mercury (the equivalent of Aristotle's moist exhalation) as the principle of fluidity (Debus 1977; Emerton 1984). Van Helmont (1579–1644), by contrast, asserted the primacy of the element "water." Readers who are interested in following the twists and turns of this story are referred to specialist texts in the history of chemistry (e.g., Debus 1977).

For our purposes, though, the family resemblances are more important than the differences between these schemes. In all of them, the major mineral classes remained stable, although explanations of their character varied according to the ontology the chemist adopted. In all of them, certain features of importance to mineralogy can be distinguished: the use of the concept of a chemical principle or element; the doctrine of mixts and aggregates; the treatment of the nature of solidity; and the choice of methods of chemical investigation. I shall examine these in turn.

All the chemists I have mentioned, as well as most of their successors until the Chemical Revolution at the end of the eighteenth century, used a small number of principles, drawn from or supplementing the original Aristotelian four, to explain the indefinitely large number of actual substances they believed to occur in the world. Each of these principles—and chemists disagreed about their nature and number, although they agreed they could not be isolated in the laboratory—conferred a certain set of properties on any naturally occurring substance in which it was present. Modify the kinds and proportions of the principles in a substance, and you altered its properties (Oldroyd 1973a). Mineralogists found the doctrine of principles congenial. One of its consequences was that the substances in the world would be indefinitely large and not marked by clear distinctions between one substance and the next. This appeared to agree with experience. With certain rare exceptions, like gold and rock crystal, minerals within a given class did not fall into clearly identifiable species, as plants or animals within a given class did. Principle chemistry offered an explanation for this.

Chemists commonly differentiated "mixts," which were chemical "compounds" of unlike parts so intimately mingled that they could not be distinguished, and "aggregates," which were associations of parts distinguishable on the macroscopic scale. Mineralogists adopted the distinction to categorize minerals and rocks. All the "simple" minerals that composed the four mineral classes were "mixts" or compounds. Minerals could also be second-order compounds—that is, principles combined with one or more mixts, or one or more mixts combined. Rocks, too, could be treated this way. Some were simply extended masses composed of one homogenous simple mineral or one second-order compound mineral; these rocks were mixts. Others were extended masses composed of distinguishable minerals; these were composite rocks or "aggregates" (Agricola 1546/1955). These distinctions were to affect both the taxonomies and the causal theories of mineralogists.

Chemical theories of consolidation were particularly important to mineralogists, since the change from fluidity to solidity served to individuate minerals and, as we shall see, to characterize major stages in the history of the earth. Changes of state were changes in the properties of bodies; chemists explained them as changes in the principles contained in a body—that is, as chemical changes. For

Aristotle, for example, water and heat were the principles of fluidity. Addition of these principles caused the transition from solid to fluid; removal caused the reverse transition.

Chemists and mineralogists distinguished two major processes of consolidation—"concretion" and "congelation." Aristotle (*Meteorologica*, book 4, chap. 6) had first described these mechanisms. A substance rendered fluid by added water (wet clay, for example) solidified (or "concreted") when the water was removed by drying or by squeezing out (the transformation of clay to bricks). A substance rendered fluid by added heat (molten glass, molten metals, or water, for example) solidified (or "congealed") when the heat was removed (the transformation to glass, solid metals, or ice). Minerals consolidated either when cold expelled their heat or when heat expelled their moisture. Metals were predominantly water, and the addition of heat caused them to "melt". Salt, soda, and dried mud, by contrast, were predominantly earth, and the addition of moisture caused them to "melt." Aristotle explained away exceptions (such as earthenware, which water did not soften) by suggesting that their pores were too small for particles of water to enter.

As chemists reconsidered the set of basic principles, they also redescribed these processes in terms of their own principles of fluidity and solidity. Avicenna picked up the Aristotelian theory of two causes of consolidation. He argued that minerals were either formed by conglutination (concretion) or by congelation. Concreted minerals were predominantly composed of the element earth, congealed minerals were predominantly composed of the element water (Emerton 1984). Paracelsus thought that salt was the principle of solidity and, consequently, that all solids contained salt. And van Helmont, in a theory that was to intrigue mineralogists for the next 150 years, thought that water could be transmuted into earth and thus change from a fluid to a solid (Debus 1977). To demonstrate this, he planted a willow sapling in a weighed amount of earth and watered it for five years. At the end of the period, he found that the willow tree's weight had greatly increased, while the earth still weighed much the same as it had at the beginning of the experiment (Webster 1966).

Mineralogists sometimes speculated about two further causes of consolidation—namely, crystallization and petrifaction—that did not seem easily assimilable to congelation or concretion. Laboratory experience suggested that crystallization took place from water. From the sixteenth century on, mineralogists accorded the process greater significance, as the widespread natural occurrence of crystals in nature was recognized. Petrifaction was a term used to describe the formation of "figured stones" (fossils).

All these different processes of consolidation produced minerals (or rocks) with characteristic textures, so that the mineralogist could use a mineral's texture to infer the mode of its formation. Glassy substances, relatively few of which were found in nature, had cooled from a melt. Stony substances, which constituted most minerals, had been deposited from water. Crystals, which were thought to be rarities, had also been deposited from aqueous solution. Petrifications were more problematic, though by the end of the century the balance of the evidence favored an organic origin. These principles of inference seemed self-evident to mineralogists until well into the eighteenth century, when experimental evidence blurred the nice matching of textures with modes of origin.

Finally, chemists regarded fire and water not just as principles but as the main agents for the laboratory investigation of minerals. Naturally occurring fire and water were, of course, only approximations to the principles, which no one expected to encounter in pure form, or to be able to isolate in the laboratory. The weaker of the two agents was water. For practical reasons, its properties, at least in its natural form, had been of intense interest since time immemorial. In antiquity, Hippocrates had emphasized the connection between water quality and health, and later Vitruvius had expounded methods for locating good water. Fire, although more mysterious, was more powerful and was the chemists' chosen agent until the seventeenth and eighteenth centuries, when water came to the fore (see chapter 3).

In the seventeenth century, J. J. Becher (1635–1682) drew all these threads together in an immensely influential cosmogony, which he published in 1669. That will form the conclusion of this chapter. But before describing it, we have to look at the rise of mechanical mineralogy and cosmogony in the years prior to Becher's work. It was the mechanical philosophers who were chiefly instrumental in introducing a genetic component to theories of the earth and in working out methods of historical reconstruction.

Mechanical Theories of Mineralogy

The mechanical philosophers of the scientific revolution discarded the ontology of principles and replaced it with one of imperceptible particles in motion. If the world really were nothing but matter in motion, as Descartes had hypothesized in the *Principles of Philosophy* (1644), then the traditional account of minerals in terms of principles or elements had to go. Instead, mineralogists had to treat minerals as "aggregates of imperceptible particles" (Steno 1669/1968, 214). Among the mechanical philosophers who took on this task (albeit as one among many more pressing ones) were Christiaan Huygens (1629–1695), Nicholas Lemery (1645–1715), Jacques Rohault (1620–1675), Nicolaus Stensen, better known as Steno (1638–1686), Robert Hooke (1635–1703), and Robert Boyle (1627–1691).

At one level, the mechanical philosophers retained many of the assumptions of the chemical mineralogists, particularly the assumptions that I have described as the common sense of the field. They continued to talk about the four main mineral classes; they thought that exhalations issued from the interior of the earth; and they, too, saw consolidation, whether by congelation or concretion, as a problem they had to explain. Of course, they reexplained the mineral classes, exhalations, and consolidation in terms of particles in motion, not chemical principles (Emerton 1984). Descartes (1644/1984, 212–13), for example, suggested that sulfur was composed of "extremely tiny and very flexible branching particles," whereas salts were "sharpened like swords." Hooke, Descartes, Steno, and other mechanical philosophers thought exhalations were responsible for earthquakes (Eyles 1958; Carozzi 1970). Indeed, Hooke (1665, preface) thought he had an instrument—the barometer—which would detect the particles of exhalations below the normal threshold of smell, such as "those steams, which seem to issue out of the Earth, and mix with the Air (and so to precipitate some *aqueous* Exhalations)." It might even be used to discover "what Minerals lye buried under the Earth, without the trouble to dig for them."

Boyle was the mechanical philosopher most concerned with consolidation.[6] He dismissed the chemists who ascribed "the Firmness and Hardness of Bodies to Salt" or some other property-conferring principle (1669, 207). Instead, according to Boyle (1669, 208), fluids were made up of particles moving at some distance from each other; solids of particles in close contact with very little freedom of movement. The three "principal causes" of solidity were "the Grossness, the quiet Contact, and the Implication [i.e., intertwining or entanglement] of the component parts."

Boyle (1669, 281) hypothesized that minerals had once been "fluid in the form either of Steams or Liquors, before their coalition and their concretion," citing as evidence the growth of fossils and minerals in rocks and the growth of stalactites in caves. But he found the chemical accounts of this fluidity too vague:

> To say with Aristotle towards the close of his third Book of Meteors, that a dry Exhalation, . . . [whether] fiery or firing, . . . makes, among other fossils, the several kinds of unfusible stones: or to tell us, according to the more received Doctrine, that Gems are made of Earth and Water finely incorporated and harden'd by cold; This, I say, is to put us off with too remote and indefinite generalities, and to found an explication upon Principles, which are partly precarious and partly insufficient, and perhaps also untrue. [1672/1972, preface].

Gems provided Boyle with his tightest arguments for the original fluidity of minerals. Their crystalline form, Boyle thought, was the result of formation in a fluid, since surrounded by a solid, they would have taken on the shape of the mold rather than their own characteristic shape. Their color, he suggested, resulted from small particles of foreign matter mixed in "when the mingled Bodies were in a fluid form" (1672/1972, 49). Their transparency showed that light could pass through them. Light, Boyle agreed with his fellow mechanical philosophers, was itself composed of corpuscles. Beams of light would be able to force their way through a fluid, pushing the corpuscles of the fluid into positions that allowed the free passage of the light. Boyle (1672/1972, 6–7) thought it unlikely "that Bodies that were never fluid should have that arrangement of their Constituent parts, that is requisite to transparency. [In a liquid] it was easie for the Beams of Light to make themselves passages every way, and dispose the solid Corpuscles after the manner requisite to the Constitution of a transparent Body."

The mechanical philosophers had a new perspective on traditional mineralogical problems that, in some cases, enabled them to shed light on problems that had previously remained obscure. Their emphasis on the shape and arrangement of the indiscernable particles assumed to be the building blocks of minerals led them to downplay the reaction of minerals to fire and water. Their emphasis on form, rather than on chemical properties, led them to treat objects with distinct forms as more amenable to analysis than the mineral classes. Hence, in the mechanical tradition, crystals, fossils, and strata received the greatest attention, whereas in the chemical tradition they had been regarded as of secondary importance to the four mineral classes.

Crystals were important because many mechanical philosophers, such as Hooke and Steno, thought that they bridged the observable world and the world of imperceptible particles (Schneer 1981; Emerton 1984). Since observable crystal shape was presumably the result of some combination of the shape and the stacking of imperceptible particles, Hooke and Steno believed they could use crystal form to make inferences about the particles that made up the crystal, a tradition that continued into the eighteenth century and beyond (Metzger 1918; Burke 1966). R. A. F. de Réaumur put forward an ambitious account of the formation of crystals in a series of three papers presented to the *Académie Royale* in 1721, 1723, and 1730. Bourguet proposed a Cartesian mineralogy in 1729, and in France the interest in the relation between crystals and the structure of matter survived the general decline of Cartesianism. Steno continued to be referred to as an authority on crystals in the second half of the eighteenth century (Romé de l'Isle 1772).

The old view that fossils had grown in place came under sustained attack by adherents of the new science, all of whom placed much more emphasis on the analogous shapes of fossils and living beings than they did on the disanalogies between the composition and locations of the two sets of bodies. Fabio Colonna (1567–1640), one of the earliest members of the Accademia dei Lincei in Naples, argued in *De glossopetris dissertatio* (1616) that fossilized sharks' teeth were organic because of their analogy to the teeth of extant sharks (Morello 1981). Steno, Hooke, and others thought that the similarity of the structure of "easy" fossils—recently petrified remains of creatures that are still extant—to known living creatures meant that fossils were the remains of living beings.[7] From this they extended the argument to more problematic fossils and claimed that they, too, were organic in origin (Schneer 1954; Greene 1959/1961; Rudwick 1972; Morello 1981). Hooke added further arguments (Carozzi 1970). Prominent among them were that fossils are composed of a wide variety of substances, unlike natural mineral species, which always consist of the same substance; that frequently the exterior of a fossil has a recognizable shape while the interior does not; and that fossils have varying degrees of hardness (Hooke 1705/1971). By the late seventeenth century the majority of mineralogists believed that fossils were the remains of living creatures. Their distribution strongly suggested that the ocean had not always been within its current boundaries. But, for reasons we shall examine in chapters 4 and 7, fossils did not become important tools of geologic inference until the early nineteenth century.

HISTORIES OF THE EARTH

The Evidence

The study of the earth's past was the mechanical philosophers' most important contribution to mineralogy. As long as the Aristotelian theory of an eternal earth held sway, doctrines of the history of the earth—so far as science was concerned—were no more than discussions of minor perturbations within an essentially stable system (Duhem 1915–1959, vol. 9). The radical critique of Aristotelian physics in the sixteenth and seventeenth centuries opened the way for an explanation of the history of the earth, as described by the testimony of the ancients, in natural philosophical terms. Physical (or natural philosophical) cosmogony flourished in the seventeenth century (Collier 1934; Greene 1959/1961; Kubrin 1967, 1968; Jacob and Lockwood 1972; Jacob 1976). Excellent descriptions of the substance of these cosmogonies abound, so I shall concentrate on the aims and the methods of their authors.

The central problem for cosmogonists was how the earth had acquired its present form and, in particular, how it had become habitable. The distinction between genetic and strictly historical theories that was later to become so prominent within geology had not yet surfaced. Most cosmogonists tried to expound general principles according to which the earth (or, for that matter, any other body in a similar position in the universe) had (or would have) developed. Thus their cosmogonies were largely genetic. Of course, things are never quite that simple in history, and there were some particular, nonrepeatable events that they recounted, most notably the Flood. But these were few and far between, and the balance of their efforts went to the formulation of laws of change.

Most cosmogonists, whether of a physical or a chemical persuasion, went about their work by the method of hypothesis. The mechanical philosophers, whose espousal of an ontology of imperceptibles had made the adoption of this method a near-necessity, found it particularly congenial (Laudan 1981). Knowing something of the present state of the earth and a little of its past, cosmogonists postulated a set of initial conditions such that, given the entities and processes of their general physics and chemistry, they could they could tell a likely story about the earth's past. Just about every physical and chemical system, including the Cartesian, Newtonian, and chemical systems, formed the base of one or more cosmogonies.

The adoption of a speculative methodology did not mean that these cosmogonists had no empirical checks. They did. These checks were the information they had about the past and present of the earth, information their cosmogonies had to fit into some coherent framework. This information we can divide into two kinds: human testimony and the "testimony of the rocks."

Cosmogonists took human testimony about the history of the earth quite seriously. The most important testimony was to be found in the first five chapters of Genesis. Most seventeenth-century scientists treated Scripture as history, as an empirical record. As such, it offered direct access to the earth's past. They regarded Genesis as one tool among others to be used in natural philosophy, not as sacrosanct (Allen 1963). Simple literalism, at least as far as those interested in the history of the earth was concerned, was already defunct in the seventeenth century. In the early part of the century, historians had debated the credibility to be ascribed to ancient legends. The consensus that emerged was that large, dramatic and public events could be transmitted by tradition, in essentials if not in detail, and that if the legends of more than one people recorded the same event, then it became a virtual certainty that the event had taken place (Momigliano 1950). Consequently seventeenth-century scholars were very impressed by the near-unanimity of most Near Eastern myths and texts by Ovid, Hesiod, and other classical authors, about the earth's early history (Woodward 1695; Ray 1692).[8]

From the point of view of the cosmogonists, the important geological themes of the early chapters of Genesis were the Creation story and the Flood. The salient features of the Creation story were the two claims that, first, the earth had originally been a watery chaos, and second, "God made the firmament, and divided the waters which were under the firmament from the waters which were above the firmament" (Genesis 1:2). Cosmogonists of all persuasions took these claims to mean that originally the earth had been fluid, and that then the solids and liquids composing the fluid had separated to form a solid surface and an underlying abyss filled with water. This water could be assimilated to the source of moist exhalations of the Aristotelian tradition, as well as providing a source for the circulation of surface water and a reservoir for the Flood.

Since there were records of a deluge or flood in the annals of almost all the literate nations then known, including Genesis, scholars confidently assumed that, at some point in the past, all or most of the earth had been submerged (Rappaport 1978). But they freely debated the extent to which this submersion had shaped the face of the earth. From the beginning of the seventeenth century, theologians of a wide spectrum of views held that the purpose of the Flood had been to show man the error of his ways. The nature of the event—whether natural or miraculous, local or universal—was immaterial. Hence, cosmogonists had considerable latitude to speculate about the origins and nature of the Flood without fear of censorship. By the late eighteenth century, most saw the Flood as simply one of a series of upheavals or "revolutions"—a term without its later catastrophic significance—that had taken place in the history of the earth.

Cosmogonists also used the "testimony of the rocks" to reconstruct the history of the earth. Since empirical knowledge of the earth, its entities, and its processes was still very limited, this only took the cosmogonists so far. They tended quickly to reduce observable phenomena to basic ontologies without the insertion of an intervening geological ontology. Insofar as they did introduce one, it tended to be the "common sense" of mineralogy.

This common sense chiefly provided evidence from minerals, rocks, strata, fossils, and veins. Chemists and natural philosophers stressed different aspects of this evidence. Chemists looked at the *composition and texture* of minerals and rocks. They constructed cosmogonies that, as we shall see, depended on differential deposition of minerals from a fluid. Natural philosophers, on the other hand, looked at the *form* of minerals and rocks. They tended to construct cosmogonies that depended on structural alterations in a cooling earth.

The Danish anatomist Steno laid out more clearly than anyone else the principles of inference that would allow the natural philosopher to infer the mode of origin, location of formation, and relative age of mineral substances (Scherz 1963, 1971; Gould 1981).[9] In his geological work, the *Prodromus* (1669/1968, 209), he explained that, "given a substance possessed of a certain figure, and produced according to the laws of nature, [he wanted] to find in the substance itself evidences disclosing the place and manner of its production." He intended to show how, given "a solid body enclosed by a process of nature within a solid," the natural philosopher could reason "from that which is perceived . . . to that which cannot be perceived" (1669/1968, 62).

His arguments all assumed that minerals had once been fluid. He first considered mode of origin or how bodies acquired their shapes. The principle he invoked was that shape depended on the ways particles had been added to a body from the fluid in which it grew.

When particles were added to the interior of the body from fluids circulating within it (and Steno believed that such percolation could occur even in the most apparently solid bodies), the growth was organic. When particles were added to the periphery of the body from fluids external to it, the growth was inorganic (220–23). In the latter case, the particles could be added in a number of different patterns, resulting in four different classes of solids within solids: deposits, incrustations, crystals ("angular bodies"), and the fossil remains of animals and plants ("replacements") (224–25).

Deposits, of which strata were the most important, were widespread layers of solids. Steno argued that they were formed when particles in a fluid fell to the bottom by their own weight, forming increments on the underlying body (227). Incrustations—noncrystalline solids surrounding a solid nucleus, such as agates and onyxes—were formed by additions made to all sides of a solid from a surrounding fluid. Crystals, such as quartz ("crystals"), hematite ("angular bodies of iron"), diamond, and marcasite, were formed when particles were added only to certain faces of an original nucleus (237–49). Finally, "replacements"—those solids within solids, such as fossils (or "figured stones" as they were often then called), which had the shape of other bodies—were formed by particles that filled a cavity left when an earlier body was destroyed (249–62). Thus, fossils (in our sense) were simply the remains of living beings, and not a special class of objects of mysterious origin.

Steno's second principle allowed him to infer the location as well as the mode of origin of mineral bodies: "If a solid substance is in every way like another solid substance, not only as regards the conditions of surface, but also as regards the inner arrangement of parts and particles, it will also be like it as regards the manner and place of production" (1669/1968, 218). Using this principle, the mineralogist could draw analogies from the known origin of minerals of a certain shape to the unknown origin of minerals of a similar shape. Since solid matter collects in layers at the bottom of turbid water, the mineralogist could infer that strata had been similarly formed. Their present location on dry land was of no matter. Since crystals found in the mountains resembled those made in the laboratory, they must have been formed by the same chemical processes. Since "bodies dug from the earth" resembled plants and animals, they must have had been formed in the same way and once been alive. Consequently, objects like the puzzling "tongue stones" of Malta were simply fossilized sharks' teeth. From at least the Middle Ages, the presence of marine shells on hills and mountains around the Mediterranean had puzzled those who studied the earth. Many scholars had already used them to argue for a relative change in position of land and sea. The arguments of Steno and others that all "figured stones" were of organic origin added further ammunition to this line of evidence about past conditions on the earth.

Steno's third principle sanctioned inferences about the relative ages of solids within solids: "If a solid body is enclosed on all sides by another solid body, of the two bodies that one first became hard which, in the mutual contact, expresses on its own surface the properties of the other surface" (1669/1968, 218). The principle of the superposition of strata, for which Steno is chiefly famous in the history of geology, is simply a special case of this more general principle. In a vertical sequence of strata, each stratum normally conforms to the shape of the top of the stratum underlying it. Thus, strata higher in the sequence have been laid down subsequently to those below them. Steno used the same principle to establish that fossils and crystals must have solidified before their matrices, and that ores and veins must have solidified after their matrices.

Once geological monuments on a scale larger than individual minerals and rocks came into the picture—monuments such as continents, mountains, and islands—cosmogonists were badly hampered by lack of information. Having listed his observations in England, Woodward (1695, 6) optimistically said, "The Circumstances of *these Things* in remoter countries were much the same with ours *here*." But the fact of the matter was that detailed observations of features of the earth's surface were few and far between. Travel was a difficult undertaking, even in the more advanced regions of Europe, and well-nigh impossible beyond Europe. Accurate mapping of most parts of the earth's surface had not yet been undertaken. Instruments for measuring the heights of mountains were only just being developed and were not to be used widely until the nineteenth century. Consequently ideas of relief were frequently exaggerated. Tenerife, for example, was generally held to be the highest mountain in the world, and estimates of its height ranged up to ten miles. No one was sure whether mountains were isolated features or arranged in chains, nor whether such chains, if they existed, were arranged in any systematic pattern. Even had accurate topographic surveys been carried out, cartographic technique was sufficiently primitive that accurate representation of relief features would have been impossible. Therefore, all that most theorists tried to explain about the major features of the earth was why there were mountains and valleys and why the face of the earth was uneven. Steno, for example, took the outline of Cartesian cosmogony and applied it to the mountains and valleys of Tuscany. To the mechanical philosophers, older nonmechanical theories that mountains grew like organic beings, or that they were the "bones" of the earth, were unacceptable (Adams 1938; Davies 1969).

Cosmogonists were somewhat better off evidentially at the largest scale, the figure of the earth. Cartesians and Newtonians might disagree about many things, but they were of one mind that the approximately spherical shape of the earth was a consequence of its formation from a rotating, fluid body (Todhunter 1873). Chemists, citing biblical evidence as the mechanical philosophers did, agreed that the earth had once been fluid. The supposed original fluidity of the earth immediately presented a problem to cosmogonists; if this were true, the earth should be perfectly spherical. By analogy with other planetary bodies the earth should "have been of an exact spherical Form, with the rest of the Earths or Planets, at the Creation of the World, before the eternal Command of the Almighty, that the Waters under the Heaven should go to their place" (Hooke 1705/1971, 313). Assuming that "in the beginning the Earth consisted for the most part of fluid Substances" (1705/1971, 325), mountains could not possibly have existed on the original earth surface.

> Some of them do seem to overhang very strangely, which cannot in any probability be imagin'd to be the form of the first Creation, it being contrary to that implanted Power of Gravity, whereby all the parts of it are held together and equally drawn towards the Center of it, and so all the parts of it ought to have been placed in their natural position which must have constituted an exact Sphere, the heaviest lowest, and lighter at the top and the water must have covered the Whole Surface of the Earth. [1705/1971, 324]

Mineralogists and geologists had no a priori reason to suspect that these arguments from human testimony, from the testimony of the rocks, and from genetic cosmogonies might lead to different and irreconcilable conclusions, forcing them to make hard choices about which was the most reliable source of information.

Physical Cosmogonies

The paradigm genetic theory within the natural philosophical tradition was propounded by Descartes in his *Principles of Philos-*

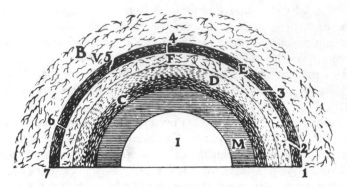

Fig. 1. The layered structure of the earth prior to the formation of mountains, according to Descartes ([1644], plate 18). *E* represents the present crust of the earth, composed of "earths" and "stones." *B* represents the air. Under the crust are layers of air (*F*) and water (*D*), corresponding to the biblical abyss (my identification, not Descartes's). The interior crust of the earth (*C*) contains "metals." Numerals 2–6 indicate fissures developing in the drying outer crust. The interior regions (*I* and *M*) were inaccessible to observation and did not concern the theorist of the earth.

ophy (1644/1984). Consistent with his ambitious program for the reformulation of natural philosophy, Descartes explained the history of the earth in terms of imperceptible particles of matter in motion. The earth had originally been a star like the sun, and hence hot. It was composed primarily, but not exclusively, of relatively large particles of matter (Descartes's "third element") and whirled around in a vortex composed of smaller particles of matter (the "second element"). In the course of time the earth gradually approached the sun, and as it did so it lost its separate vortex. At the beginning of its approach, the earth comprised three concentric shells. The interior region consisted of tiny particles of the "first element," aetherial matter; the surrounding shell was composed of particles of the first element welded together; and the outermost shell was made up of particles of the third element interspersed with those of the second.

Once the earth had descended into the sun's vortex, the outer shell split into five different layers, separating the different mineral classes (fig. 1). Earths and stones formed the present exterior of the earth, the fourth layer out from the center. Air (large particles of the third element interspersed with small particles of the second element) made up the layers above and below the earths and stones. Under the inner layer of air was a layer of water (corresponding to the biblical abyss in other cosmogonies), made up of long, smooth particles. Metals, including their ores, formed the interior crust of

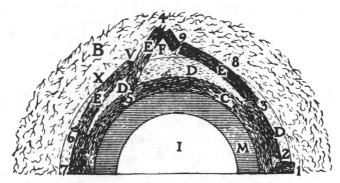

Fig. 2. The formation of mountains and oceans, according to Descartes ([1644], plate 18). Portions of the exterior crust separated by fissures (4–5) have fallen by their own weight, creating mountains (4–9), allowing the water under the crust to escape to form oceans (2–3 and 6–7) and leaving some level plains (VX).

the earth. Their solidity was caused by the branching, intertwined particles of which they were made up.

Mountains, plains, and oceans had been formed when parts of the earthy exterior crust of the globe fell into the waters beneath (fig. 2). The water under the earth's surface supplied the rivers, oceans, clouds, and rain. "Sweet water," made up of slender and flexible particles, could circulate through cavities in the fallen exterior, whereas the hard particles of salty water could not. Fresh water rose to the surface and into the air as vapor, only to fall as rain when the small particles of air could no longer support it. Earthquakes occurred when exhalations, collected and ignited in the earth's cavities, suddenly rarefied causing the earth to shake or even to erupt in a volcano.

Some physical cosmogonists, such as Steno, followed Descartes's model closely. Others chose different models. Whiston (1696), for example, based his theory on Newtonian, not Cartesian, physics and utilized the periodic approach of a comet that upset the globe's stability. Woodward (1695) explained the history of the earth chiefly in terms of gravitational forces. The physical cosmogonists were handicapped not only by their very inadequate data but by the ambiguity of reconstructing a history of the earth from physical principles. Each individual physical system could be conjoined to so many different initial conditions, and these were so arbitrarily chosen that the constraints hardly excluded any proposed system. Furthermore, the cosmogonists could choose from a variety of competing physical systems. Even so, they succeeded in introducing genetic and historical components to the study of the earth that had only had an embryonic presence before, but that were never again to disappear; and they succeeded in combining these components with a commitment to empiricism.

Chemical Cosmogonies

Even more influential in the development of mineralogy and geology than the cosmogonies constructed by the natural philosophers were those put forward by the chemists. Eighteenth-century mineralogists looked back to J. J. Becher's *Physica subterranea*, published in 1668, as the source of this tradition (Oldroyd 1974; Debus 1977; Ashworth 1984; Emerton 1984).[10] (fig. 3). Becher's

Fig. 3. The frontispiece to Becher's *Physica Subterranea* (1668). Becher shows the globe composed of the three elements—earth, water, and air—as well as the three kingdoms of nature—the animal, the vegetable, and the mineral.

work is confusing, not to say inconsistent, at times, and any exegesis is likely to impose a clarity not present in the original. But perhaps this need not matter for our purpose, which is less to examine Becher in detail than to give the outlines of an approach to mineralogy and cosmogony that later scholars found suggestive. Becher began with a description of the creation of the heavens and the earth, a description that, naturally enough, relied heavily on the biblical account. But the bulk of the work was devoted to the production of solid bodies out of the fluid chaos that was the original earth and the subsequent differentiation of those bodies to form the earth that we now observe, including its live inhabitants.

This was where chemistry, rather than the Bible, began to play a part. Becher postulated chemical processes, such as crystallization and precipitation, as well as the old doctrine of exhalations, as the means by which solids had separated from the fluid chaos. Like most chemists of the time, Becher was an eclectic, favoring the preservation of snippets of earlier works if they still seemed to have some value, so that he combined Aristotelian element theory with strands drawn from Stoicism and from Paracelsianism (Emerton 1984). He drew an analogy between two "laboratories" in the world, that of nature—the subterranean—and that of art—the superterranean. Becher himself shows little sign of having systematically used practical experience in the laboratory of art to illuminate the subterranean processes in the laboratory of nature that led to the formation of the present earth. His chemical theory appears to have been largely a priori. But in the generation or two that followed him, the use of analogies drawn from practical laboratory experience to construct genetic theories of the earth's origin became more and more prominent. Consequently, the empirical control on cosmogonies became much greater than when they had been constructed primarily by the method of hypothesis.

Becher identified three chemical principles, "earth," "water," and "air." For Becher, earth—the most important of the three—came in three forms: "vitrifiable," "fatty," and "mercurial." The vitrifiable earth was the principle of fusibility, solidity, diaphaneity, and opacity; it was equivalent to Paracelsus's salt. The fatty earth was the principle of color, taste, and smell; it was equivalent to Paracelsus's sulfur. The mercurial earth was the principle of malleability and ductility; it was equivalent to Paracelsus's mercury. All three earths were to be found in solution in the waters that Becher, following the biblical account of the abyss, believed to lie under the solid exterior of the earth.

For Becher, as for later chemical cosmogonists, the separation of water and earth described in the first chapter of Genesis was more important than the Flood, and the explanation of the various kinds of mineral substances found on its surface was more important than the explanation of the larger-scale features such as mountains and oceans. Hence he gave lengthy descriptions of the generation of the four main mineral classes in the abyss, which provided a watery matrix to supplement Aristotle's earthy matrix (Oldroyd 1973a). Here the liquid embodiment of the first earth—universal acid, or spirit of vitriol—reacted with smaller amounts of the other earths to form the mineral classes of metals, earths and stones, salts and sulfurs. His characterization of different species within these classes closely followed the earlier taxonomies of Agricola and Boethius de Boodt. Salts and bituminous minerals (as well as stars, animals, and plants) were formed by a combination of water and earth. Stones, metals, and earths were formed of combinations of different earthy principles. Stones grew from a matrix of vitrifiable earth, earths from a matrix of fatty earth, and metals from mercurial earth.

The circulating waters of the abyss transported the minerals to the veins and lodes in which they were found on the earth (Emerton 1984). In many cases, they were deposited as crystals from the watery solution. The bulk of the solid earth, in which these minerals were deposited, was the vitrifiable (siliceous) earth, widely observable in impure form as sand, quartz, and rock crystal. Becher concluded with a lengthy discussion of the subsequent decomposition and breakdown of the mineral kingdom.

The impact of Becher's theory was enhanced by Georg Stahl's adoption and adaption of his chemistry. Stahl identified Becher's fatty earth with sulfur, named it "phlogiston," and elevated it to a major role in chemistry (Oldroyd 1973a). Since Stahl taught most of the next generation of mineralogists in his medical and chemical courses at Halle, he was responsible for transmitting a version of Becherian mineralogy and cosmogony even though he himself (despite an essay he wrote in 1700 on the origin of mineral veins) was less interested in mineralogy than Becher had been. The Becher-Stahl tradition dominated speculation in Germany, Sweden, and to a lesser extent France for the next couple of generations about the history of the earth. That will form the subject of the next chapter.

3
The Becher-Stahl School of Mineralogy and Cosmogony, 1700–1780

Chemical mineralogy and cosmogony flourished in the eighteenth century. This tradition has received much less scholarly attention than the tradition of physical cosmogonies, yet it was the tradition that gave rise to Werner's theory of the earth, and hence ultimately to historical geology. In the century between Becher and Werner, chemists and mineralogists systematically used the method of analogy to construct a genetic, causal account of the earth's past based on laboratory experiments. At the same time, the institutional setting in which they worked—state mining bureaucracies and training schools—directed their attention to the field as well as to the laboratory. Consequently, by the last quarter of the eighteenth century, the empirical base of mineralogy and cosmogony had been substantially broadened, and cosmogonies, while remaining genetic, embodied increasingly detailed information about the structure and history of the earth.

THE INSTITUTIONAL SUPPORT FOR CHEMICAL MINERALOGY

In the eighteenth century, continental European states commonly owned and operated a variety of industrial enterprises. These institutions were one result of a prevailing mercantilist ethos that encouraged rulers to undertake actions to enhance the wealth of the state (Baumgärtel 1963). Some state industries had no connection with mineralogy—textile manufacture, for instance. But a surprising number of them, particularly the mining, metallurgical, and porcelain industries, fostered the study of mineralogy and geology.

Mining, in particular, supplied a utilitarian motive for the study of minerals (Porter 1981). Mining officials and instructors had to face practical issues, unlike the members of royal academies who, whatever the intentions of their patrons, tended to revert to high theory whenever possible.[1] They needed information about the location and properties of metallic ores such as lead, zinc, copper, tin, and, most important of all, silver. Silver formed the bulk of the money supply in Europe in a period when a subsistence economy was giving way to a money economy, and it was much in demand.

Over the course of the century, state mining officials interested themselves in a growing variety of minerals besides the metallic ores. Kaolin (the raw material for the growing porcelain industry), rock salt, limestone (to burn for quicklime), stone or clay (to make bricks), sand (for glassmaking), rock (for road building), and lime and sand (for smelting) were much sought after. Demand for these valuable raw materials and for a better understanding of their properties proceeded in tandem with the industrialization of Europe.

The various German states—particularly Saxony, which had access to the rich ore deposits in the Erz Mountains—led the way in the bureaucratization of mining (Baumgärtel 1963). The explosive growth of silver mining in that area in the late fifteenth and early sixteenth centuries stimulated the foundation of new towns such as Freiberg, Joachimstahl, and Chemnitz. Although there was a slump in the 1530s caused by the opening of the Spanish silver mines in the New World, the abundant mineral resources of the region ensured a steady trickle of openings for mineralogists, first in the state mining bureaucracies, and later in state mining schools. The region around Freiberg was the busiest and most productive of all mining areas, even if far removed from the intellectual centers of the day.

In turn, Saxon mining officials played a disproportionately large role in the development of mineralogy and geology during the next couple of hundred years. Agricola (1494–1555) spent his adult life as physician and town official in the boom mining towns of Joachim-stahl and Chemnitz in the Erzgebirge. Balthzar Roesler (d. 1673), author of *Speculum Metallurgie* (1700), Carl Friedrich Zimmermann (d. 1747), author of *Obersächsische Bergakademie* (1740), and F. W. von Oppel, author of *Anleitung zur Markscheidekunst* (*Subterranean Geometry*) (1749), held high positions in the Saxon mining service. J. G. Baumer, who wrote *Geographia et Hydrographia Subterranea* (1779) was a counselor of mines, and Friedrich Wilhelm Hein von Treba (1740–1819), whose essay on the interior of mountains was published in 1785, was subdirector of mines. Wilhelm Charpentier (1738–1805) and Abraham Gottlob Werner taught at the Mining Academy at Freiberg. Friedrich Hoffmann (1660–1742), a distinguished chemist and author of *Dissertatio de Matricibus Metallorum* (1738), worked in the mining service in Freiberg and Naples.

Mining officials from other regions also contributed to the growth of mineralogy and chemical cosmogony. Johann Lehmann (1719–1767), originally trained as a physician, as were many chemist-mineralogists, was in charge of copper production in Prussia during his most productive years. He ended his career as professor of chemistry in St. Petersburg. Georg Füchsel (1722–1773), author of a treatise on the rocks of Thuringia, was a royal physician, but he was also keenly interested in mining operations in his native state. Giovanni Arduino (1714–1795) was inspector of mines for Tuscany before becoming professor of mining and metallurgy in Venice. Emanuel Swedenborg (1688–1772), who was a serious mineralogist before turning to more mystical pursuits, and Axel Cronstedt (1722–1765) both worked for the Swedish Bergskollegium, or Board of Mines. This had been founded in the mid-seventeenth century to administer the mining industry, but it also came to sponsor research in mineralogy and metallurgy (Frängsmayr 1974).

Following the introduction of high-quality Chinese porcelain into Europe in the sixteenth century, European rulers embarked on a costly and competitive quest for the secret of its manufacture, a search that provided further jobs for mineralogists. In 1693, the French set up the first works at St. Cloud outside Paris. Others followed in Vienna, Höchst, and Nymphenberg, as well as Berlin and Meissen. They were substantial enterprises; by mid-century Meissen and Berlin, for example, each employed about 400 people. Mineralogists soon realized that kaolin, or china clay, was the crucial ingredient, and they began to search for deposits of the newly precious material across Europe (Gillispie 1980). By 1709, the Meissen works, founded by Frederick Augustus I of Saxony, had successfully produced porcelain following the discover by J. F. Böttger (1682–1719) that certain fluxes (alabaster, marble, or feldspar) rendered the clay fusible. This remained a closely guarded state secret, so that Frederick the Great instructed the chemist J. H. Pott (1692–1777) to try to solve the puzzle independently. Pott, a pupil of Hoffmann and Stahl, proceeded systematically, reportedly making over 30,000 different experiments subjecting different substances to intense heat in the furnace. He did not discover the secret of porcelain manufacture, and Frederick the Great passed him over for election to the Berlin Academy. But he did publish his results, which included one of the first experimental distinctions of different earths (Pott 1746; see also Pollard 1981).

All these state industrial enterprises, particularly the mining ones, provided bureaucratic positions for mineralogists. Initially the positions were filled by anyone with prima facie plausible qualifications, frequently chemists trained in medical schools. But agitation soon arose for specialized training to meet this need. German reformers saw the universities as outmoded medieval institutions. They argued that, rather than trying to attempt the probably impossible task of updating the universities, separate institutions for professional education should be established (Turner 1974). The technical schools that were founded in the German states, and their imitators elsewhere, were staffed by Stahlian chemists, a procedure that in turn ensured that mining officials and other bureaucrats in the state industries saw the world through Stahlian spectacles. Johann Henckel (1678–1744), a near-contemporary of Stahl's, was the bridge between Stahlian chemistry and the mineralogy of the mining academies. Trained in medicine by Stahl, Henckel spent the 1710s and 1720s as a physician in the mining town of Freiberg. For the last decade of his life he served in the Saxon mining service, first in Dresden and then in Freiberg, with the task of surveying Saxony's mineral wealth. In what spare time he could find, he taught metallurgical chemistry privately (Herrmann 1962). His pupils included A. S. Marggraf (1709–1782), who was appointed to the Berlin Academy (over Pott) on the basis of his important and extensive work on salts, metals, and minerals in solution, and J. A. Heynitz, one of the chief architects of the Bergakademie in Freiberg. Johann Cramer (1710–1777), who was successively a councillor for the Hapsburg and Saxon mining administrations, and Christlieb Gellert (1713–1795) similarly taught metallurgical chemistry on an informal basis. Thus informal training in subjects related to mining—mineralogy and metallurgical chemistry—was available by the mid-eighteenth century.

Despite these informal efforts in the 1730s, governments remained unimpressed by the call for more formal instruction. A proposal put to the Saxon government laying out the organization and utility of a mining academy was rejected (Hufbauer 1982). Lehmann's advocacy of specialized technological institutions met with no more success (Freyberg 1955).

After the Seven Years War, all this changed. The travel reports of the French chemist, mineralogist, and mining engineer Antoine Gabriel Jars (1732–1769) may have been a precipitating factor. Most were first published in the *Mémoires* of the Paris Academy of Science, and subsequently compiled by his brother. In 1757, the French government had sent Jars with another chemist, Henri Duhamel (1700–1780), to examine the mines in Saxony and in the Austrian Empire (Chevalier 1947). Their report revealed how little

Hapsburg mining officials knew about the operations they supposedly controlled, and it reinvigorated the movement for technical training. In 1765 the mining academy at Freiberg in Saxony was formally established (Herrmann 1953). Gellert became the first professor of metallurgical chemistry (Baumgärtel 1961). Shortly thereafter, the Hapsburg Empire recognized the academy at Chemnitz; and in 1770, C. A. Gerhard (1738–1821), a Prussian mining official, organized the Berlin Mining School.

In Sweden, the reform went somewhat differently, for the universities created positions that served technical interests. In 1750, the mineralogist Johan Gottschalk Wallerius received the first chair of chemistry at Uppsala. He was succeeded in 1767 by Torbern Bergman (1735–1784).

Impressed by German and Swedish technical know-how, the French prudently interested themselves first in their theories, and then in their institutions. In the 1750s and 1760s, Rouelle (1703–1770), the master teacher and leading French chemist, acted as the conduit through which Stahlian chemical mineralogy, already commonplace in Germany, reached the French (Rappaport 1960). His pupils, including Lavoisier and Desmarest, as well as other French mineralogists such as Guettard, were sympathetic to the Stahlian chemical program he espoused. As part of his crusade, Rouelle enlisted the help of Baron d'Holbach (1723–1789). D'Holbach, a German-born naturalized Frenchman, disseminated useful technical information in every way he could, particularly in the compilation of the *Encyclopédie*. But more important for our purposes, he translated into French Wallerius's *Minéralogie* (1753), Henckel's *Introduction à la minéralogie* (1756), C. E. Gellert's *Chimie métallurgique* (1758), a selection of J. G. Lehmann's collected works under the omnibus title *Traités de physique, d'histoire naturelle, de minéralogie et de métallurgie* (1759), and Stahl's *Traité du soufre* (1766) and *Traité des sels* (1771). Added to translations by other chemists, this meant that the French had ready access to works in the German-Swedish tradition.

This sustained campaign to improve French mineralogy paid off. In 1783, the French founded the École des Mines in Paris, in the face of determined resistance from such working geologists as Antoine Monnet (1734–1817), who thought that mining should be learned underground and not in a school (Aguillon 1889; Birembaut 1964; Gillispie 1980; Taylor 1981–82). The École thrived, and its graduates found jobs in the Corps des Mines. Though the Corps numbered only in the twenties and was thus an order of magnitude smaller than

the Corps associated with the École des Ponts et Chaussées, it
provided employment for mineralogists. The associated journal, the
Journal des Mines, founded in 1792, flourished immediately, offering an outlet for specialized papers on mineralogical and geological subjects.

In one area, the French outstripped the Germans and the
Swedes—namely, the inventorying and representation of the mineral resources of the state in the form of a mineralogical map. The
French *Atlas et Description minéralogiques de la France*, published
in 1780, is the highest achievement in this genre (Rappaport 1969b;
see also Eyles 1972). Its main architect was Jean-Étienne Guettard
(1715–1786), who had presented a preliminary mineralogical survey
of France to the Académie Royale in 1746. His attempts to revise
and refine this came to the attention of Henri Bertin, minister and
secretary of state in charge of mining. In 1766, Bertin, perhaps
mindful of Jars's findings on the support for mining enterprises in
Germany, provided some financial assistance for Guettard. As a
result, the young Lavoisier, who became an accomplished geologist,
joined the project (Rappaport 1968, 1969a, 1973). Still the survey
dragged. In 1777, Antoine Monnet, who was already an inspector of
mines, took over, promising Bertin quick results. Three years later
he published the *Atlas*, or at least 31 quadrangles out of a projected
total of 214.

Guettard and Lavoisier (though not Monnet, who was deter-
minedly atheoretical) were both sympathetic to the Stahlian chemical mineralogy then becoming popular in France, yet their theoretical interests were not reflected in their maps. This was conscious,
if not uncontroversial, policy. Guettard stressed that the purpose of
the *Atlas* was to show the location of minerals, building stone, and
agricultural and industrial soils, and to help civil engineers find
road-building materials in unfamiliar areas of France (Guettard
1751). He thought that representing general geological relations was
unnecessary and inappropriate on a map of mineral deposits, even
though he believed that the country could be divided into three
concentric *bandes* of rocks, one of sandstones and limestones, one
of marl, and one of metallic ores, schists, and granites. Consequently, he represented mineral resources just as he would have
done towns and villages, by spot symbols dotted across the map (fig.
4). He resisted Lavoisier's call to show the areal extent of deposits,
though Lavoisier succeeded in including some sections in the *Atlas*
showing idealized vertical relations between the rocks.

To summarize, a reinforcing cycle developed in many countries.

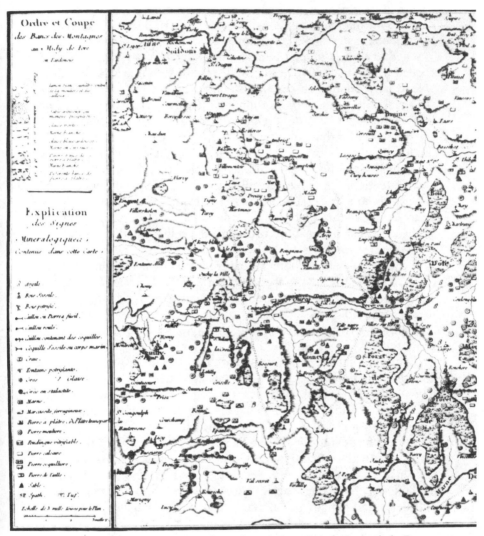

Fig. 4. Part of a sheet from the *Atlas et Description minéralogiques de la France* (Guettard and Monnet 1780), entitled "Une partie du Valois, Soissonois, Rhemois et pays adjacents." This plate was drawn up by Guettard and Lavoisier. Deposits of minerals, many of them of economic significance, are indicated by spot symbols. The general distribution of the underlying rocks is not shown.

Institutions were set up to supply bureaucrats for state industries, and the graduates of these institutions could count on employment. Naturally those who paid the pipers presumed that the tune would encourage the growth of industry. They expected mining officials to

ensure steady production, and teachers in mining schools to concentrate on mining techniques and the utilitarian aspects of metallurgical chemistry and mineralogy.

Whether or not mining officials and their teachers actually succeeded in the task of fostering industrial development is questionable. Many eighteenth-century scientists, such as Monnet, were
skeptical (Gillispie 1980). The mining officials did not directly profit
from increased mine output. Consequently, they did not necessarily
see profit, rather than job security or the production of high-quality
products, as the *sine qua non* that it was rapidly becoming in the
individualistic industry of Britain. Some of the countries with the
best educational systems—Sweden, for example—continued to be
backward industrially (Pollard 1981). Indeed, it may be anachronistic to see the increase of wealth, rather than the increase of control,
as the aim of the bureaucratization of the mining and metallurgical
industries.

The uncertain effects of technical education on economic progress
largely resulted from the inability of theoretical science to clarify
mining and metallurgical technologies. Generations of practice, not
theory, ensured success. A glance at the major phases of mining and
metallurgy—prospecting, mining, assaying, and smelting—will
show this. Prospecting, whether for minerals that occurred in ore
deposits and that were the focus of attention on the Continent, or for
minerals, such as coal, that occurred as strata and that were the
focus of attention in England, remained pragmatic, largely uninflu-
enced by scientific theory, at least into the middle of the nineteenth
century (Torrens 1978, 1984). Geological knowledge sometimes
narrowed the search for new ore deposits—for example, by postulating a rough correlation between mountainous areas and ore
deposits. But generally, prospectors relied on the time-honored
methods of following surface deposits underground, examining
hillsides after rain to see if unusual deposits had been exposed,
tasting spring water for unusual metallic tastes, running trial sink-
ings and borings, and (most dubiously) using the divining rod
(Gough 1930; Duckham 1970).

Theory shed little light on mining. Practical problems predominated: how to shore up the shafts, tunnels, and adits constituting the
mine; how to ensure ventilation and drainage; how to move men and
materials in and out of the mine; and how to prevent explosions.
Assayers, by contrast, relied on some knowledge of mineral properties, though again they had usually acquired this by hands-on
experience rather than by systematic investigation or theory. A

partial exception was Cronstedt's demonstration in the middle of the century that the blowpipe could be used to distinguish an impressively large range of different minerals (Oldroyd 1974b). Metal smelting and working similarly owed more to generations of experience than to any codified body of knowledge (Smith 1960). Therefore, the systematic mineralogy taught in the mining academies (as opposed to surveys of existing techniques) probably did little to stimulate industrial development in Europe.

However, the converse was true: industrial development stimulated theoretical mineralogy by providing jobs in mining schools and mining services. Although the pursuit of systematic mineralogy was not the primary focus of mining school training, it steadily gained in importance. An increasing proportion of instruction in mineralogy and geology accompanied the more practical aspects of the curriculum. For example, at Freiberg, Werner taught theoretical courses such as mineralogy, geognosy (first offered in 1778 and covering much the same ground as an introductory geology course), and paleontology (first offered in 1799), as well as practical courses such as ferrous metallurgy, mineral geography, the history of Saxon mining, the preparation of mining reports, and mining economics (Ospovat 1971). Courses of the former kind served to codify, propagate, and preserve the geological and mineralogical knowledge that was generated in eighteenth-century Europe. Science thus benefited from institutionalization, achieving both continuity and a cadre of individuals with the time and inclination to pursue mineralogical and geological problems. European states had constructed a system in which a number of scientists—admittedly small by our standards, but nonetheless significant in terms of the science of the day—were financially secure and had sufficient leisure to contemplate theoretical as well as practical matters. Scholars traveled to the leading mining academies, and the international exchange of ideas became easier.

The contrast with Britain is instructive. Both the nature and the relative value of British mineral resources encouraged the isolation of British mining engineers and mineralogists from continental practice. Although there were areas of Britain where metal ores were mined—the Cornish tin mines and the lead mines of the Mendips and Derbyshire—the British had a much greater interest in "stratiform" deposits such as coal, in large measure because of the shortage of wood on the island (Gough 1930; Burt 1969). The techniques for prospecting for and handling coal were different from those for metal ores; as a result, owners and managers in these industries had little cause to emulate continental practice. There was, however, considerable interest among those who worked with metal ores, particularly in Cornwall, and a regular passage of personnel back and forth from Freiberg.

British mineralogists did not enjoy the luxury of training in mining schools and guaranteed government employment. Mine managers were trained on the job (Pollard 1965). Porter has argued that in Britain there was no clear connection between the industrial revolution and the rise of geology, except in the case of a few rather isolated "practical men" like William Smith, who had relatively little impact on the overall patterns of development (Porter 1973). Practical men thoroughly understood the nature and distribution of minerals in their particular area, but their excessive localism and lack of access to modes of publication inhibited their ability to reach the wider learned community (Torrens 1978, 1981). Eighteenth-century British mineralogical practice lagged behind its continental counterpart, and it is hard to resist the conclusion that at least part of the reason was the lack of even haphazard institutional support. As a time when it was still not easy to publish, the isolated amateur or practical man had little opportunity to pass on his results in the absence of any institutional setting. British mineralogy depended on the whims of the wealthy or of professionals with enough leisure to devote themselves to learned pursuits. And mineralogy was not high on the list of favored British eighteenth-century leisure activities.

ADVANCES IN MINERALOGY

What was the impact of institutionalization on the development of mineralogy and geology? I shall concentrate on four areas in which I believe the impact is both clear and significant, mentioning others in passing. They are (1) mineralogists' increasing interest in and knowledge of *rocks* as opposed to minerals; (2) their refinement of the description of the ocean presumed to have deposited these rocks; (3) their increasing ability to distinguish different minerals, particularly the earths, by chemical means; and (4) their increasing comprehension of the role of heat, as well as water, in consolidation. These advances are reflected in the more detailed chemical cosmogonies produced in the eighteenth century.

The Rocks

For reasons we examined in the last chapter, mineralogists had traditionally treated rocks as chemical "mixts"—that is, large masses of one mineral —or as "aggregates"—that is, as congeries of individual minerals distinguishable by eye. As a result, rocks had scarcely seemed worthy of study in their own right, for once there was a theory of minerals, a theory of rocks could be assumed to follow automatically.

But this was to change, for mineralogists employed in mining services had to explain features of the rocks that were not readily reducible to mineralogy. How were mineralogists to distinguish different rocks? Was mineralogy always adequate to this task? How could they correlate rocks in one area with those in another? What was the relationship between ore-bearing rocks and topographic features? Why were valuable ore minerals so irregularly distributed across the earth's surface? Why did they occur in certain rocks and not in others? And so the queries continued.

By the second half of the eighteenth century, mineralogists had formulated some widely accepted answers to these questions, particularly those about the distribution of the rocks. A number of different mineralogists argued that the rocks could be divided into a small number of distinct groups. "Primary" rocks were flanked by the stratified, gently sloping "secondary" rocks that made up gentler terrain. (fig. 5). The primary rocks were usually granitic, surrounded by schists or gneisses, and banked-up layered sediments. They formed the cores of mountain chains and were frequently criss-crossed by veins containing valuable ore minerals.

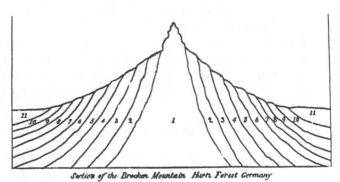

Section of the Brocken Mountain Hartz Forest Germany

Fig. 5. The late eighteenth-century theory of mountain structure (Philips 1816, 115). Drawn over half a century after Moro, Bertrand, and Lehmann first proposed this general account of mountain structure, Phillips's diagram shows how long the account persisted. According to Phillips, the structure of the Brocken mountain in the Hartz was typical of all mountain structure. The numbers refer to Werner's identification of the sequence of rocks. The core (1) is granite, surrounded by (2) clay-slate; then come the transition rocks, (3) limestone and (4) greywacke, followed by the secondary or *Flötz* rocks, (4–10) sandstones, limestones, and gypsum, and finally (11) the alluvial deposits.

Secondary rocks contained petrifactions (fossils), unlike the primary rocks.

Who exactly was the first to describe rocks in this way, or whether it was a case of simultaneous, independent discovery, is a tangled question, and not one that I shall try to answer here.[2] The Italian cosmogonist Lazzaro Moro (1687–1764) distinguished primary and secondary rocks in his *De' crostacei e degli altri corpi marini che si trovano su' monti* (1740), which was translated into German in 1751. The Swiss geologist Élie Bertrand proposed three different classes of rock in his *Mémoires sur la structure intérieure de la terre* (1752; also see Carozzi and Carozzi 1984). Lehmann admired this work, and put forward a similar account (Lehmann 1756). He distinguished two basic classes of rock—the *Ganggebürge*, which contained veins of ore, and the *Flötzgebürge*, which did not. In the late 1750s, Rouelle differentiated the *terre ancienne*, composed of massive, granitic, unstratified deposits, from the *terre nouvelle*, composed of approximately horizontal layers of stratified rocks (Rappaport 1960). Similar distinctions were advanced by Pierre-Simon Pallas (1741–1811), Giovanni Arduino, and Georg Füchsel.

In all these classifications of rock, elements besides the purely mineralogical crept in—considerations of location, elevation, dip, fossil content, age, and mode of origin. And connected to this new awareness of the geography of rocks was the appearance of sections and maps to show their relationships. Mine sections had been commonplace since the late seventeenth century, though for economic reasons they usually remained private. Far more important were the traverse sections across ten, twenty, or more miles of country in which the geognosist pictured what might be seen if the landscape were sliced vertically (Rudwick 1976; Taylor 1985).

The Ocean

The growing popularity of spas and health cures in the eighteenth century accentuated chemists' and mineralogists' concern with aqueous solutions. They wrote more papers on mineral springs than on any other chemical or mineralogical topic (Homer Le Grand, personal conversation, 1983). Much of this was, of course, busy-work, undertaken on commission for spas and resorts in a period when chemists were glad to find any paying job. Wallerius, for example, had spent a spell as superintendant of Danemarks, a spa near Uppsala, where his duty was to analyze the spring water. But busywork or not, mineralogists accumulated an impressive body of information about the wide range of substance dissolved in mineral waters, including ocean water. Thus, speculations about what might have been dissolved or suspended in the watery abyss gave way to evidence about what was actually dissolved in naturally occurring waters. As information about the diversity of natural waters accumulated, mineralogists briefly considered the possiblity that water might not be a single element but a number of different elements. In his volume on hydrology, the *Hydrologia ell Wattu-Rikket* (1748), Wallerius tried to classify the different kinds of water. In the end, however, he concluded that there was only one kind of water.

New evidence for a diminishing ocean, over and above the supposed testimony of Moses, accumulated. In the 1690s, the Swedes had observed that, relative to the land surface, the water in the Baltic had fallen. This stimulated a whole flurry of speculations. Swedenborg suggested the change in level was caused by the decreasing speed of the globe's rotation. Celsius (1701–1744) adopted a variant of Whiston's theory whereby the effect was attributed to the passage of a comet. Wallerius resorted to the time-honored, if dated, view that this was yet one more example of the conversion of water into earth (Wegmann 1969). But whatever the cause, their belief in the reality of the effect—a lowering of sea level—was reinforced. This suggested a cause for the distribution of the secondary rocks that flanked the cores of mountains.

The Differentiation of the Earths

Traditionally, as we have seen, chemists and mineralogists had paid less attention to the "earths" than to the other major mineral classes—metals, salts, and sulfurs. Consequently, they had little to say about the mineralogy of most rocks, since these were primarily earths. In the eighteenth century that changed. Chemists differentiated a number of earths, abandoning the old idea that earth was a single element. Mineralogists identified rocks corresponding to these different earths. New accounts of the mineralogy of the rocks, unlike new accounts of their structure and distribution, owed more to the laboratory than to the field. Almost without exception, the chemists and mineralogists who contributed to the mineralogy of the earths accepted Stahl's approach to chemistry. They included Henckel, Hoffmann, Lehmann, Woodward, Pott, Gellert, Wallerius, Cronstedt, and Bergman (Emerton 1984).[3] So, too, did some important figures we have not yet mentioned, such as the French chemist Pierre Macquer (1718–1784), Balthazar-Georges Sage (1740–1824), founder of the École des Mines, the Swedish naturalist Linnaeus, and the crystallographers Romé de l'Isle and René Just Haüy (1743–1822). The Stahlian tradition was gradually modified over time, and by the mid-eighteenth century it often had a strong admixture of Cartesianism (in the case, say, of the crystallographers) or Newtonianism (in the case of Bergman). Eighteenth-century chemistry, like earlier chemistry is complex, and I shall examine only those portions of it that bore on mineralogy.

For our purposes, the work on the nature and variety of the earths was by far the most important. Becher and Stahl, it will be remembered, had postulated three earths, but only the first of these corresponded to earth in the sense of the mineral classes of earth, metal, salt, and sulfur. During the eighteenth century, Stahl's second earth (sulfur or phlogiston) lost its earthy nature and became solely a principle, and his third earth (mercury) vanished altogether. Over time, then, the vitrifiable earth advanced from the first of three earths to the sole earth principle. No sooner had this happened than chemists broke it down into a whole sequence of earths, each of which they assumed to be elementary.

In his influential chemical compendium, Macquer had concluded that "the most general and probable opinion is, that as only one kind of fire, of air, and of water, so only one kind of simple elementary earth exists" (Macquer 1766/1777, article on "earths"). He recognized that there appeared to be different kinds of mineral earths, which he described in some detail, but maintained that these differences resulted from impurities. The purest earth was characterized by its inertness; it was the material that remained following the destructive distillation of a substance by heat. If strongly heated itself, "earth" fused to form a glass, and hence was often termed "vitrifiable," a term we have already encountered.

The mineral that most nearly approached the principle was widely identified with rock crystal (quartz or silica). Many mistakenly assumed diamond to be the purest form of rock crystal (Emerton 1984). But there were a whole family of minerals, including (in modern terms) quartz, chalcedony, flint, agate, the micas, amphiboles, pyroxenes, olivine, feldspars, and clays, that were formed largely of silica or its cognates. This nicely reinforced the increasing knowledge of the order of the rocks, since the major component of the primary rocks was "vitrifiable earth."

But by the time that Macquer had identified a single earth, the belief that there was one and only one basic earth was being called into question. A major reason for this derived from new experimental techniques. By the second half of the century, Cronstedt (1770) confidently claimed that the laboratory was the ultimate tribunal for the distinction of minerals. Mineral properties were best studied in the laboratory, either by heating them or by dissolving them in water.

The use of heat for mineral analysis had a long history. By the beginning of the eighteenth century, mineral assayers could estimate with some accuracy the amount of a metal that could be recovered from an ore on the basis of its behavior when heated. Heat also served to distinguish some varieties of earth. For a long time, mineralogists had recognized that one variety of earth—the "calcareous earth"—was not resistant to heat, but turned to a powder or "calx" on heating. And some twenty years before Macquer wrote, Pott distinguished the calcareous earth from the vitrifiable and gypsous earths as a result of having heated several thousand substances, primarily earths, to find the secret of porcelain manufacture (Pott 1746).

But it was the action of aqueous fluids, not of heat, that proved the key to distinguishing the other earths. The use of heat had always been somewhat suspect. Although heat broke down more substances than water, the status of the products was unclear. Since they usually assumed heat to be a substance in its own right, chemists wondered if the products of heating were not new substances, rather than the components of the substance being analyzed. Like Rouelle and others, Bergman (1783, 10) was scathing about the method, which, he thought, "however useful in other respects, rather tend[s] to confound than to lay open the component parts of bodies." Chemists and mineralogists gradually shifted to an almost exclusive use of the "wet way," refining techniques for investigating the interactions of acids, bases, and salts in solution (Holmes 1971).

For geologists the earths were pivotal, even though, for chemists, the salts may have been of greater interest. Mineralogists distinguished most earths in the "wet way." Friedrich Hoffmann described magnesia, J. G. Gahn (1745–1818) and C. W. Scheele (1742–1786) recognized "terra ponderosa" or "barytes," and A. S. Marggraf described argillaceous earth (Oldroyd 1973b). Analysis in the wet way showed that the earths were partially soluble in naturally occurring water. By the 1790s, mineralogists recognized that even silica (vitrifiable earth) was slightly soluble in water and in basic solutions. Bergman, Klaproth, Kirwan (1733–1812), Dolomieu

(1750–1801), Guyton de Morveau (1737–1816), and Joseph Black analyzed both liquids and solids, helping elucidate silica solution (Newcomb 1985). Black, for example, studied water that the president of the Royal Society, Sir Joseph Banks, had brought back from the hot springs of Iceland. The springs had appeared to contain siliceous petrifactions, unlike the calcareous petrifactions commonly found in Europe. Black determined the water's silica content by saturating the alkaline mineral waters with acid, evaporating part of the solution, and then allowing a jellylike precipitate to form over a period of weeks. He found between 2.1 and 3.75 grains of silica per 1,000 grains of mineral water (Black 1794). As a result of such laboratory investigations of naturally occurring waters, mineralogists had good reason to believe that siliceous (vitrifiable) earth was soluble at least in certain circumstances. Chemical cosmogonists who hypothesized that the crystalline rocks, including siliceous rocks, had once been dissolved in water had good empirical grounds for their claim.

By 1784, mineralogists knew enough about the behavior of the earths in solution for Bergman to publish a synoptic scheme for mineral identification. The basic trick was to bring the mineral into solution by fusion with alkali, then to perform a series of precipitations and resolutions, similar to those performed today in analytic chemistry (Oldroyd 1973b). After Bergman, Nicholas Vauquelin (1763–1829) in France and Richard Kirwan continued to refine the method, which came to its culmination with the work of the German chemist M. H. Klaproth (1743–1817).

When the English chemist James Keir (1735–1820) translated Macquer's *Dictionary* (1766/1777, footnote to article on "earths"), he added a note explaining how Macquer's belief that there was only one kind of elementary earth, which was siliceous or vitrifiable, was out of line with contemporary practice:

> The class of earth called vitrifiable contains many substances, the properties of which are exceedingly different, [and] no proof can be given that those substances, which are believed to be the purest of this class of earths, as diamonds, are more simple and elementary than calcined calcareous earths, argillaceous earths, earth of magnesia, metallic or many other earths; all which, when purified by ordinary methods, are as incapable of decomposition as diamonds are, and consequently as well entitled to be considered as elementary.

Joseph Black (1803, 1:342–43) announced in his lectures (published posthumously), "Of the purest and most simple earthy matter, six or more different kinds are now reckoned, which we cannot reduce to greater simplicity by our operations." These six were the calcareous earth, the argillaceous earth, the siliceous earth, the gypsous earth, the barytic earth, and the magnesian earth. All of them, except the barytic earth, formed major rock types—namely limestone, clays and shales, sandstones, gypsum, and dolomite. The significance of the distinction of the earths for the Chemical Revolution has been recognized for some time (Siegfried and Dobbs 1968). It was just as significant for geology; mineralogists could use it to assess different theories of the origin of the rocks made of these minerals. In the late eighteenth century, the balance of the evidence favored formation from water.

Finally, we should note that Cronstedt laid one lingering mineralogical problem to rest. By a series of arguments and experiments, Cronstedt (1770, xiii) established that there was no chemical difference between earths and stones, only a difference in hardness or "induration." Earths and stones "consist[ed] of the same principles" and were "by turns converted into one another." Since chemistry now ruled, mineralogists took this as decisive.

The Causes of Consolidation

As we have seen, mineralogists had agreed that rocks and minerals consolidated in four different ways: concretion, congelation, crystallization, and petrifaction. Furthermore, the consensus was that the first, third, and fourth ways occurred from aqueous solutions. Laboratory work on solutions confirmed these conclusions. At the descriptive level, mineralogists' account of consolidation, just like their account of the mineral classes, remained stable. At the theoretical level, they reinterpreted the phenomena. Richard Kirwan, for example, described consolidation in Newtonian terms— that is, in terms of the forces between the particles he assumed composed minerals.

Kirwan (1799, 107–8) argued that the cohesive power came from two sources "the *general* attraction, or gravitation" and the short-range forces beloved of chemists—"the *specific* attraction of the integrant particles of one species of matter, either to each other, or to those of another species." Although he thought that "exact contact perhaps never takes place," nonetheless "both these sorts of attraction are so much the stronger, and, consequently, so is the resulting hardness, as the points approaching to contact are more numerous and nearer to each other in the same mass, and these are capable of becoming so much the more numerous the particles that present them are more minutely divided."

"Concreted" rocks, such as clays, shales, and other argillaceous rocks, were formed by "the close union of earthy particles to each other, without any sort of crystallization, but arising merely from their approximation to each other after the expulsion of the superfluous water" (Kirwan 1799, 124). They still contained a small portion of water, which prevented them from crumbling to dust; concreted rocks with more than a small portion of water were soft, like the rocks in the interior of mountains.

Kirwan granted that crystals could grow from melts. Crystallization occurred, according to Kirwan (1799, 109-10), when the "component particles of bodies are so arranged in uniting to each other, as to assume a regular internal, and external form," which can be achieved only when the bodies are "minutely divided, have liberty of motion, . . . [are] placed at a due distance from each other, and [are] undisturbed by a force superior to that of their mutual attraction." Such minute division occurred either in "igneous solution" or in "solution in a liquid menstruum." Hence, crystallization could take place either in the wet way or in the dry way. Nonetheless, Kirwan downplayed the importance of crystallization from melts; the particles were so crowded that good crystals of, say, quartz or calcareous spar would be unlikely to be formed. "Cemented" rocks, such as breccias and sandstones, were formed by the introduction of particles between larger chunks (Kirwan 1799), and "petrifactions" (fossils) were formed by the replacement of organic matter by stony or metallic matter.

But if one development in eighteenth-century mineralogy was the reinterpretation of older categories of consolidation, there was also a contrary tendency at work. Laboratory investigations into the behavior of heated earths began to break down the classic distinctions between stoniness, glassiness, and crystallinity, the states presumed to be the result of concretion, congelation, and crystallization, respectively. Previously, mineralogists had explained these different textures in terms of different modes of origin, and ultimately of the presence of different chemical principles. Stony substances (by far the commonest) had precipitated from water; glassy substances had cooled and congealed from melts; and crystalline substances, containing a saline principle, had grown in aqueous solutions.

But when René-Antoine Réaumur (1681–1757) and Jean Darcet (1725–1801) experimented on the behavior of earths in porcelain ovens, they found that a fused earth could be restored to a stony, granular condition (Réaumur 1727, 1729; Darcet 1766; Smith 1969). For some considerable time this attracted little attention among geologists, who continued to assume until the early nineteenth century that stony substances were deposited from aqueous solution (see chapter 6). Furthermore, Réaumur as well as Pierre-Clement Grignon, Guyton de Morveau, and Jean-André Mongez suggested metals that had melted could cool to form crystals (Taylor 1975). James Keir (1776), a glass manufacturer and a competent chemist, showed that slowly cooled glass could crystallize. This, he claimed, was evidence against a chemical theory of principles, and for a Newtonian theory of matter, since the contrast between a glass and a crystal was not the result of the presence of different principles, but of the arrangement of the constituent particles. Lavoisier argued a similar case. The state of a body, he claimed, whether solid, liquid, or elastic depended on the quantity of the heat substance (which he was later to call caloric) that it contained. Congelation and crystallization were essentially the same phenomenon, not the radically different processes that the Stahlian chemists had claimed. Crystals could grow from either heat or water as a solvent (Taylor 1975). If these conclusions were correct, then neither stoniness nor crystallinity was an unequivocal guide to the mode of formation of a rock.

Some mineralogists applied this new information to the question of the origin of rocks without delay. Keir, for example, argued that the spectacular columns of basalt found at the Giant's Causeway in Northern Ireland and Fingal's Cave in Scotland were large crystals that had formed from a melt, not from water. But it was to be some time before these chemical conclusions came to play an *important* role in the theory of the earth. One reason may have been that crystals and glasses, which were the chief substances whose origin was under reexamination, were relatively rare. Everyone was prepared to admit that glasses were formed by heat, but they were normally found in the area of volcanoes. Crystals were still not recognized as forming many rocks, nor were they to be until rocks were examined in thin section in the nineteenth century. And as for stones, the recognition that most of these that could be readily deposited from water may have delayed the realization that they could also form from melts. At the beginning of the nineteenth century, James Hall, a Huttonian, continued to have to produce experimental evidence for the consolidation of earths and stones from melts. In chapter 6 we shall examine the use that Hutton made of these suggestions.

Eighteenth-Century Chemical Cosmogonies

Chemical analysis in the wet way formed the empirical basis of geology by the mid-eighteenth century. Chemist-mineralogists assumed that the processes by which solid earth separated from the waters of the ocean were analogous to laboratory reactions. Where mechanical philosophers and seventeenth-century chemists had relied almost exclusively on the method of hypothesis to construct their cosmogonies, eighteenth-century mineralogists used the method of analogy to move from laboratory experience to the minerals they found in the world. That is, they knew that in the laboratory certain effects were linked to certain causes. For example, crystals grew in aqueous solutions only when those solutions were undisturbed. Chemists observed analogous effects in the world—crystalline minerals, for example. By analogy, they inferred that naturally occurring crystalline minerals must have grown in an undisturbed aqueous solution.

Cosmogonies in the Becher-Stahl tradition were predicated on the deposition of the rocks of the earth's crust from water (Oldroyd 1974a). As knowledge of the respective solubilities of the different earths increased, mineralogists applied this knowledge to cosmogony, assuming that the rocks had been deposited according to their degree of solubility. Field evidence supported this view, for everyone was agreed that the primarily siliceous rocks (rocks made of "vitrifiable earth") underlay (and hence had presumably been deposited prior to) the clayey (argillaceous) and calcareous rocks, and these in turn underlay the rock salts and gypsums. The French geologist Déodat de Dolomieu, for example, proposed that each of the chemist's simple earths (lime, magnesia, argillaceous earth, and siliceous earth) had been deposited in different epochs of the earth's history, depending on their solubility. Over the century, Becher's promissory notes about a cosmogony constructed using evidence from the laboratory were made good. The cosmogonies were, however, still largely genetic, less a reconstruction of individual events than an outline of general causal laws of the earth's development.

Some features of older cosmogonies dropped out. Chemical cosmogonists said rather little about exhalations. They were divided about the presence of an abyss. Some regarded it as a reservoir for water; others, such as Werner, denied its existence. Other topics received greater attention. For example, the theory of mineral veins became a central concern because of the connection of the mineralogists with the mining industry.

Most of the chemical mineralogists whose work has been mentioned in this chapter proposed cosmogonies based on their mineralogies. J. F. Henckel, J. H. Pott, J. G. Lehmann, J. G. Wallerius, T. O. Bergman, G. F. Rouelle, and A. Baumé were only some of the more prominent. I shall mention only two or three of these men, and even them very briefly. Chapter 5 will be devoted to a lengthy discussion of the culmination of this tradition in the work of Abraham Gottlob Werner.

Henckel, like Becher and Stahl, applied chemical theory to cosmogony in his voluminous publications on mineralogy in the period between 1720 and 1740 (Henckel 1725, 1722, 1747; Oldroyd 1974a). He believed that dry land had originally separated from water, as Moses and other ancient writers had described. The original ocean was supposedly a thick, coagulated, and gelatinous fluid, quite unlike the waters of the present ocean. Henckel allowed that explanation of consolidation in terms of particles rather than chemical principles might make sense; he suggested, for example, that crystals are formed by the accretion of minute earthy corpuscles. Such eclectic combinations of chemical and mechanical accounts of the basic entities were commonplace in the eighteenth century.

J. G. Lehmann accepted Becher's threefold division of the earths and paid particular attention to the formation of metals, which he thought was brought about by adjusting the quantities of the three elementary earths in a compound ("mixt") under the agency of air and water (Lehmann 1756). Like Henckel, Lehmann blended Becher's theory of the earth with somewhat more mechanical modes of explanation. The appropriate entities of the construction of a chemical cosmogony were minerals. Even more important, Lehmann integrated this with a discussion of the sequence of primary and secondary rocks.

Bergman was probably indebted to Lehmann (Schufle 1967). In 1773–74, after reading Lehmann, Bergman published the second edition of his *Physisk Beskrifning Öfver Jordklotet* (*Physical Description of the Earth*), which had originally appeared in 1766. He followed Lehmann in believing that rocks, as well as minerals, had to be used to reconstruct the earth's history. In this work, which was quickly translated into German, he divided the rocks into primary, secondary, and "accidental" (Frängsmayr 1969).

Bergman believed that most rocks had been laid down in water, which had once covered the whole earth. Since the water's composition had varied from time to time, and from place to place, the rocks it deposited naturally varied also. The primitive rocks made up the earth's core and were probably as old as the earth itself. Most were various forms of vitrifiable earth—quartzites and granites, for example—although some were calcareous earth in the form of crystalline limestones. They had been deposited from solution or from very fine suspension, as water was lost from the earth, presumably by evaporation. None contained fossils. Next came the stratified rocks. They were usually found as gently sloping deposits on the flanks of the primitive mountains. Bergman believed them to have been derived largely from the submarine erosion of the primitive rocks. They were stony, not crystalline, suggesting that they had been deposited from suspension, not solution. Most of them were formed of more soluble earths—the calcareous and gypsous earths, for example, in the form of limestone and gypsum—though some vitrifiable earth still occurred as sandstone. The stratified rocks were abundantly fossiliferous. Finally, there were two other "accidental" rocks of more recent origin, the alluvial and volcanic rocks. The former had been deposited in very shallow and localized bodies of water; the latter had been ejected from the earth's interior fire as ashes, stones, and lava.

By the second half of the eighteenth century, then, mineralogists perceived a nice fit between their genetic causal theories of earth development and the testimony of the rocks they observed in the field. The former, based on the analogy of chemical experiments carried out in the laboratory, suggested that rocks (which were nothing more than extended masses of minerals, primarily earths) should have been deposited in the inverse order of their solubility. The latter showed that, indeed, the primary rocks were composed largely of siliceous earths, and the secondary, stratified rocks of more soluble earths.

EIGHTEENTH-CENTURY PHYSICAL COSMOGONIES

On the continent, physical cosmogony continued to flourish.[4] Treatises by Lazzaro Moro in 1740 and Benoît de Maillet (1656–1738) in 1748 and the belated appearance of Leibniz's cosmogony, the *Protogea*, in 1749, kept the tradition alive. But increasingly, as the century wore on, even those whose primary allegiance was to the physical tradition, had to take the new findings about minerals into account.

The dominant figure was Buffon (1707–1788) (Roger 1963). Although his cosmogonies fall outside the main thread of my narrative, they are worth a brief mention since they show how physical cosmogonists were taking increasing account of chemistry and mineralogy. In the *Théorie de la terre* (1749), Buffon had combined a cosmogony based on the familiar concept of a cooling earth with a geology that employed the action of water on rocks originally formed by fire as the main causal agent. This was probably his most influential work (Taylor 1968). By 1778, when he published the *Époques de la nature*, Buffon had come to the conclusion that heat played a major role in the late as well as early stages of earth history (Roger 1962). In the volumes of mineralogy of the *Histoire naturelle* (which Buffon, in collaboration with Daubenton, published rather late in his life, from 1783 on), he proposed an interesting synthesis of cosmogony and mineralogy. He argued that minerals should be arranged according to the order in which they had been formed, rather than according to some static hierarchy supposedly imposed by God. The mind best grasps ideas, he insisted, when they occur in the succession in which they were ordered in time (Buffon 1778). The "matières brutes et minérals" that made up the earth could be divided into three great classes, distinguished by the order in which they appeared on the earth. This in turn determined both their location and chemistry.

The first class comprised all those substances that Buffon thought had been produced by the "primitive fire" from which the globe had cooled. These rocks formed the interior of the globe and also jutted out to the surface in places. They were solid, glassy "vitrifiable" rocks, though unlike his contemporaries, Buffon attributed their origin to fire, not water. The second class of rocks, primarily lavas and basalts, was formed by the further action of this fire, particularly in volcanoes. Neither of the first two classes contained fossils. The third class, which included the calcareous and vegetable earths, was formed by the action of water on rocks of the first two classes. Here, at last, the remains of organized beings were to be found.

Roger (1973, 23) has argued that theories of the earth, including Buffon's, constituted the "intellectual frame of the sciences of the earth from the end of the XVIIth to the beginning of the XIXth century, [and were] not a preliminary sketch to modern geology." This observation is sound so far as *physical* cosmogonies are concerned, but the case is quite otherwise with chemical cosmogonies. Few mineralogists, or chemists interested in mineralogy, paid any attention to Buffon.

4

The Botanical Model Rejected

In the previous two chapters we have looked at the causal genetic theories advanced by mineralogists and cosmogonists and at the categories of minerals and rocks that they used in these theories. But of equal or greater concern to mineralogists was the status of the various minerals. Were earths, metals, salts, and bitumens natural kinds? How were these natural kinds related one to another? If the answer to the first question was negative, then these classes ought not play a role in causal theories. If no answer was forthcoming to the second question, then mineralogists had to admit they had no grip on the structure of the mineral kingdom.

These are the kinds of questions that taxonomists address. Eighteenth-century taxonomists, I shall argue, believed that their job was to sort out the diversity in the world, to identify natural kinds, and to specify the relations between them. The taxonomic urge was much more fundamental than simply a desire to develop pedagogic aids, as has sometimes been suggested (Porter 1977). Classification required both theory and experience. Theory was needed in order to specify which characters were essential to natural kinds—a decision that could not be made by simply looking at a succession of animals, plants, or minerals. It was also needed in order to assess whether or not a correct classification had been constructed, for this assessment was based on the classification's specification of kinds that could function in successful scientific theories. And of course, experience was needed, for only by consulting the world could candidates for natural kinds be identified, and only by putting classifications to the test of experience could their adequacy be assessed.

On this interpretation of the nature and function of taxonomy, it should come as no surprise that the construction of mineral classifications was an abiding concern, almost an obsession, with mineralogists. New classifications appeared almost annually during the eighteenth century.

In accepting such an interpretation of classification, I realize that

I am treading on tricky ground; in the popular mind, as well as in the minds of many historians and philosophers, classifications are merely conventional organizational schemes, comparable to the arrangement of goods on grocery store shelves or books in a library. This is not the place to argue at length for my interpretation, because philosophers of biology have already advanced compelling arguments against the popular interpretation, at least so far as the natural history sciences are concerned (Hull 1974; Rosenberg 1985).

The major figure in eighteenth-century systematics was Linnaeus. He achieved his greatest success in botany. In the years following the appearance of Linnaeus's *Systema Naturae* (1735), Swedish and German mineralogists applied the concepts and methods that had worked so well in botany to the taxonomy of minerals, just as they extended chemical theory to causal processes involving minerals. Discouragement set in as a whole generation of mineralogists concluded that the mineral kingdom and the plant kingdom differed so radically that the concepts and methods that worked for the latter could not be applied to the former.

As a result, most mineralogists (except those in France) welcomed Abraham Gottlob Werner's proposal that the practical task of effectively and reliably identifying minerals be given precedence over the more theoretical project of classifying them. The significance of this proposal will become clearer in chapters 5 and 7, but is worth mentioning now in order to motivate some of the more technical parts of the discussion in this chapter. Werner's disillusionment with the applicability of traditional canons of taxonomy to the mineral kingdom meant that he was willing to contemplate classifications that flouted some of these canons. When he came to classify the rocks, he allowed that their essence might be the age at which they were formed, thus simultaneously flying in the face of traditional practice and opening the way to a strictly historical geology.

This was no trivial achievement. Historians of biology have long recognized that traditional taxonomic practice impeded an evolutionary account of species. That same practice impeded a historical account of the earth. The "discovery of time" was not delayed simply by religious prejudice but by some of the most deeply rooted philosophical principles in the Western tradition.

THE BOTANICAL MODEL

Eighteenth-century taxonomy was dominated by the Swedish naturalist Linnaeus (1707–1778) (Cain 1958; Larson 1971; Mayr 1982;

Frängsmayr 1983a). In the last decade or so, scholars have realized just how complex Linnaeus's views actually were. But for our purposes, all that is important is the eighteenth-century perception of his taxonomic principles, even if that perception was skewed in significant ways.

Most eighteenth-century naturalists believed that Linnaeus was committed to the following positions: that the world was made up of "natural kinds"; that these had to be classified by their "essential characters"; that there was some "subordination of characters" such that taxonomists could proceed by considering only one or a small number of characters, rather than the complete set; and that the natural kinds in the world could be grouped into a hierarchy of higher taxonomic units—namely genera, classes, and families. All these claims are *assumptions* about the structure of the world, and hence fallible. Botanists and zoologists found that they worked fairly well for the plant and animal kingdoms; mineralogists found many of them inadequate or misleading in the case of the mineral kingdom.

The aim of taxonomy, Linnaeus believed, was to expose the order that God had imposed on nature. "We should study the works of nature" he said, because it is the "business of the thinking being, to look forward to the purposes of all things; and to remember that the end of creation is, that God may be glorified in all his works" (Linnaeus 1735/1802, 1:1). A true classification would reflect the order of God's creation (Lindroth 1983). If botanists followed his "method," they would find that "amidst the greatest apparent confusion, the greatest order is visible" (Linnaeus 1735/1802, 1:3). Thus, Linnaeus believed that ideally classifications should reflect the world and transcend mere conventions. In practice, of course, the natural historian, could not always achieve this lofty goal.

To construct his theory of natural kinds, Linnaeus drew on Aristotle's distinction between essences and accidents of a substance. Andrea Cesalpino had already applied these distinctions to botany, though his use of Aristotelian logic went well beyond anything Aristotle himself had envisioned (Sloan 1972). From Cesalpino, Linnaeus took the view that the taxonomist aimed to discover the essence of the kinds being classified—that is, the property or properties that made the substance what it was. All other properties were "accidental" in the sense that they could be present or absent without altering the identity of the substance. Furthermore, Linnaeus accepted that the characters of substances were arranged in such a way that a single character, or a small number of characters, could be used to determine this essence, and that all other characters were subordinate to it or them.

In the case of plants, these characters were the reproductive organs, or so Linnaeus asserted (Gunnar 1983). The doctrine that reproductive systems were the essence of plants had an honorable, though controversial, history going all the way back to Aristotle. Linnaeus set to one side the residual doubts about whether plants reproduced sexually that some botanists still harbored in the 1720s, since he believed that God had organized the world that way. Sexual reproduction guaranteed that, with only unimportant variations, plants stayed stable from generation to generation. Furthermore, the sexual organs of plants were easy to observe, individuate, and count. Thus, they served for the identification and classification of most plants.

Finally, Linnaeus believed that God had imposed a hierarchical order on the world. He had created individuals, species, and genera in systematic relationships. Thus, the objects in the natural world could be arranged in "five branches, each subordinate to the other: *class*, *order*, *genus*, *species*, and *variety*," although he was more confident about the reality of the lower taxa than of the higher ones (1735/1802, 1:1). Unlike Cesalpino, Linnaeus fixed the terms *species* and *genus* at specific ranks in the hierarchy, in principle, if not always in practice. At first, Linnaeus (1735/1802, 1:3) had advocated that every natural body should be given its own unique name, which would "point out whatever the industry of man had been able to discover concerning [the body]." For convenience, he contracted this to two names for each species of plant, that of the genus to which it belonged, and its own species name. This binomial nomenclature so effectively reduced terminological confusion in botany that it became one of the hallmarks of Linnaeus's method and was widely adopted in mineralogy and chemistry as well.

In spite of scathing criticisms from other naturalists, most notably Buffon, Linnaeus used these principles to bring order into the chaos of rival schemes of botanical classification. Other disciplines tried to follow his example. But before looking at those who tried to apply Linnaean taxonomy to mineralogy, we should look at his own attempts.

THE BOTANICAL MODEL APPLIED

Linnaeus tried to apply the taxonomic principles he had worked out for botany to mineralogy (St. Clair 1966; Oldroyd 1974d; Engelhardt

1980). He rejected the two major mineralogical traditions that we have already examined, arguing that he intended to replace the "*Physical*, which descends through the obscure generation of minerals . . . [and the] *Chemical* which ascends through their destructive analysis" in favor of the "*Natural*, which considers their superficial and visible structure" (Linnaeus 1735/1802, 7:9). To show that mineralogy could be treated in the same way as botany and zoology, Linnaeus had to establish that minerals were natural kinds, that their essences mapped on to observable characters, preferably of a generative or sexual nature, and that the mineral kinds could be hierarchically arranged.

Linnaeus had reservations about whether the first of these conditions was fulfilled; he was not at all sure that there were mineral species analogous to plant species, since he was not sure that there was any analogue in the mineral kingdom to the sexual reproduction that ensured stability of plant kinds. Linnaeus (1735/1802; 7:9) lamented that the "calamity" of mineralogy was that it was "scarcely ever, that the Species can be sufficiently determined, since in [minerals] the generation proceeds not from the egg; but [from] irregularly sportive nature." Like many of his predecessors, he was well aware of the practical difficulties of individuating mineral species, and he suspected that many minerals were no more than varieties.

However, he attempted to pursue some slight parallels between the generation of plants and the generation of minerals by accentuating the nascent sexual imagery in the familiar Becherian cosmogony. Beginning from the conventional premise that the globe had once been fluid, primarily the element or principle "water," Linnaeus (1735/1802, 7:3) proposed that the water had generated a "double offspring . . . a saline male and a terrene female." The saline male (salt or acid) was the form-giving principle, impressing shape on the terrene female (earth). Therefore, salts and earths were the "fathers" and "mothers" of minerals proper.

Linnaeus postulated four different salts and four different earths, one each for the atmosphere, the ocean, plants, and animals. The salts were "nitre" (the "aerial" salt), "muria" (the marine salt), "natrum" (the animal salt), and "alum" (the vegetable salt). The earths were sand (crystallized from rainwater), clay (precipitated from seawater), soil (from vegetables), and calx (from animals). These earths and salts combined in various ways to form Linnaeus's variant of the now-familiar mineral classes. Combinations of earth (as mixts) formed the *petrae*—the earths and stones—which

Linnaeus subdivided into the vitrifiable, calcareous, and refractory. Combinations of earths and salts formed the *minerae*—the salts, sulfurs, and metals. Combinations of earth (as aggregates) formed the *fossilia*—petrifactions, concretions, and earthy soil.[1]

The reader will notice that the sexual metaphor took Linnaeus only a very limited way in his classification. Indeed, it could be argued that it played no real role in the classification of the actual mineral kinds found in the world, as opposed to the chemical kinds out of which Linnaeus assumed they were generated. For actual minerals, Linnaeus used either crystal form or chemical behavior. Crystal form had long been assumed to be a reflection of the essence of those minerals that exhibited it (see chapter 2). Following this tradition—and consistent with his own requirement that the taxonomist use observable, easily individuated, and easily counted characters—Linnaeus proposed that the mineralogist classify crystalline minerals by counting their various faces. Although admirably consistent, Linnaeus's strategy of using crystal form had the distinct disadvantage that very few minerals are crystalline to the naked eye.

For the noncrystalline minerals, Linnaeus turned to the equally well-established tradition of treating the chemical composition of minerals, as revealed by their reactions to heat and to water, as their essential character. Hence, he divided the earths and stones into those that fused, crumbled, or remained unaffected by heating. The one significant exception was the petrifaction group, which, as the remains of living beings, Linnaeus classified by shape. Later in this chapter, we shall see how mineralogists in the generations after Linnaeus continued to differ about whether mineral form or mineral composition should be given taxonomic primacy.

Again, in considering the relation of the various mineral groups one to another, Linnaeus tried to extend the principles he had already worked out for botany. He set up a complete hierarchy of mineral kinds, ranging from orders down to classes, genera, species, and varieties. His urge for order was so great that, as we have seen, he even made this hierarchy symmetrical, dividing each of the three orders into three subdivisions and so on down the hierarchy.

Very few, if any, mineralogists accepted Linnaeus's mineral classification *tout court*. But his methods had proved so successful in botany, and his assertion of an analogy between the plant kingdom and the mineral kingdom seemed sufficiently plausible, that, for the next half century, mineral taxonomists' theory and practice was shaped by their reaction to the Linnaean system.

THE BOTANICAL MODEL EXTENDED

The French, with their commitment to the Cartesian heritage, were much more sympathetic to the formal aspects of the Linnaean system—the use of crystal form for classification and the insistence on a hierarchy of mineral classes—than their German and Swedish counterparts. I shall indicate only the outlines of French mineralogy, which remained tangential to developments within geology until well into the nineteenth century.

The three key figures are Romé de l'Isle, René-Just Haüy, and Déodat de Dolomieu (Hookyaas 1952a, 1953, 1955; Burke 1966; and esp. Mauskopf 1970, 1976). Romé de l'Isle was struck by the compatability between Linnaeus's insistence on the importance of crystal form and the mechanical theory of the microstructure of crystals. Although Romé de l'Isle had once thought that a saline principle caused crystal form, and hence that variety of chemical composition caused the variety of crystal forms, he later assimilated his chemical theory to a corpuscular one. Crystals, he suggested, were composed of small saline *"molécules intégrantes,"* themselves composed of acidic and alkaline *"molécules constituantes."* Each mineral had a fixed structure and a fixed composition; hence, the mineral classes so necessary for Linnaean taxonomy were assured.

In taking this stand, the French mineralogists were turning their backs on Buffon, who had sharply criticized Linnaeus's insistence on the reality and fixedness of species. Romé de l'Isle and Haüy had both been introduced to mineralogy by Daubenton (1716–1800), Buffon's protégé and sometime collaborator. In a long and somewhat uneasy relationship, Daubenton had assisted Buffon with much of the *Histoire naturelle*, although not with the mineralogical volumes. Even so, in his own mineralogical text (1784) and in his lectures at the Jardin du Roi, Daubenton accepted Buffon's contention that there were no species in the mineral kingdom. Daubenton's lectures prompted Haüy to turn to mineralogy from his first love, botany. Very possibly it was his botanical training that inclined Haüy to argue against his mentor and for the existence of mineral species. Haüy (and Dolomieu, too) proposed that Romé de l'Isle's *molécule intégrante* corresponded to the individual plant. The combination of *molécules intégrantes* in regularly shaped, chemically homogenous crystals corresponded to the plant species (fig. 6).

But problems continued to haunt this school of mineral taxonomy. The old difficulty that relatively few minerals appeared to be crystalline persisted. Before polarizing microscopes were used to

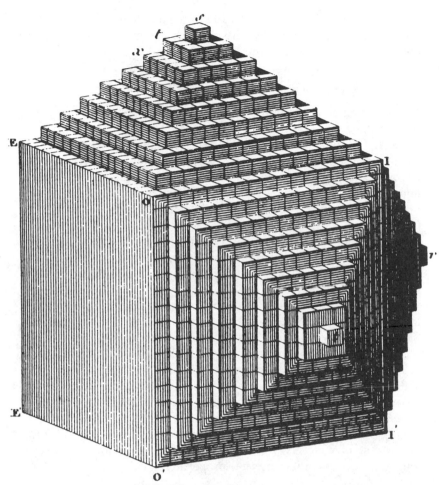

Fig. 6. Haüy's diagram of the formation of a rhombic dodecahedron by the progressive decrement of "integrant molecules" on the edges of layers added to a cubic nucleus. He attempted to establish analogies between plant individuals and species and mineral "integrant molecules" and crystals and thus to put mineral taxonomy on a stable basis. (Haüy 1822, atlas, plate 5).

study thin sections—a development that did not occur until the 1830s—mineralogists contended primarily with minerals that appeared non-crystalline. Furthermore, crystal form, even when evident, did not always neatly correlate with chemical composition. One substance—calcium carbonate (calcareous earth), for example—could appear in more than one crystal form (aragonite and calcite). Conversely, one crystal form could be exhibited by substances of different chemical compositions (the different alums).

Only at the end of the second decade of the nineteenth century were these conundrums settled by Mitscherlich's theory of isomorphism and polymorphism (see chapter 8).

THE BOTANICAL MODEL REJECTED

We now turn to the majority tradition within mineralogy, concentrating particularly on Axel Cronstedt, Torbern Bergman, and Abraham Gottlob Werner. All of them thought that, whatever else was to be said about minerals, their essential characters were chemical. As a consequence, all had trouble with the notion of fixed mineral species (Werner 1774/1962). Chemists, after all, believed that there were only a small number of "elements" or "principles." Combined in various proportions, they comprised the variety of substances actually found in nature. This led the chemical mineralogist to suspect that there was an indefinitely large number of indistinct mineral varieties, not a finite number of distinct mineral species.

Furthermore, if the essence of minerals was chemical, then their taxonomy could not be based on external characters. Cronstedt thought that crystal form was no way to classify minerals. He protested that his taxonomy was an attempt to "obtain some protection against those who are so possessed with the *figuromania*, and *so addicted to the surface of things*, that they are shocked at the boldness of calling a *Marble* a *Limestone*, and *of placing the Porphyry among the Saxa*" (Cronstedt 1758/1770, xxi).

Torbern Bergman (1783, 6–7) also rejected the use of crystal form in the case of minerals. Form could be used for plants because it reflected their essential, internal properties.

> There is a power implanted by the Creator in organized bodies, which, upon the acquisition of proper nutriment, unfolds and evolves the structure which before lay concealed in the fecundated egg or seed. . . . Hence it is that the leading features on the external parts agree with the internal properties, and when judiciously chosen, form sufficient characteristic distinctions.

But, in Bergman's mind, form did not reflect the essential, internal properties of minerals. "The formation of fossils [minerals]" said Bergman (1783, i), "is totally different. Here no system of vessels collects, distributes, secretes or changes the concurrent particles, but they run together by chance, and are solely connected by the power of attraction." Consequently, minerals should be classified by their "constituent or component parts." These could be determined only by chemical analysis, not by casual examination.

"In methodizing fossils," Bergman suggested that "compounds should rank under the most abundant ingredient," measured by relative weight. In practice, he ran into difficulties following his own advice. To produce an intuitively satisfactory classification, he had to make all kinds of exceptions to his rule. If the mineral's ingredients were "not of the same intensity," then the ingredients of greater intensity were to count more in determining the genus, though how intensity was to be determined he did not say. If the mineral were economically valuable, he added, "the value of the thing must . . . be considered." "Minerals containing gold and silver must be ranked with those noble metals although they hold three, four or more times the quantity of heterogenous matter" (Bergman 1783, 10).

Cronstedt and Bergman were divided on the question of whether mineral kinds could be arranged hierarchically. Cronstedt (1758/1770, xii) inveighed against the very idea. He intended his system "as a bar or opposition to those, who imagine it to be an easy matter to invent a method in this science, and who, *entirely taken up with the surface of things*, think that the *Mineral Kingdom may with the same facility be reduced into classes, genera, and species, as animals and vegetables are.*" The reason was that in the plant and animal kingdoms each individual had a single, characteristic form; minerals, on the other hand, often exhibited several, detectable only by chemical analysis. "In the last two kingdoms of nature," he said, "there are but seldom, and never more than two different forms found mixed together in one body, whereas in the mineral kingdom it is very common, though it will nevertheless always remain concealed from anyone, however penetrating, *who has not employed himself in the compounding or decompounding such bodies.*" Bergman, by contrast, accepted Linnaeus's taxonomic hierarchy without demur, and did not question that minerals could be divided up into classes, genera, species, and varieties.

But Linnaeus's most forceful and influential critic was Werner. Since Werner looms so large in the history of mineralogy and geology, a short biographical introduction is in order. He came from a family that had long been associated with the mining and metallurgical industry, and after several years' experience in an ironworks, he enrolled in the recently founded mining academy in Freiberg in 1769. There he decided to enter the Saxon mining service and, since a law degree was a prerequisite, moved to Leipzig two years later. His interest in mineralogy (and in other subjects such as the history of language) proved greater than his dedication to law, and he left Leipzig in 1774 without taking his degree. However, during his time in Leipzig, Werner had translated a mineralogical work by J. K. Gehler, *De characteribus fossilium externis* (1757), as well as studying other mineralogical treatises. In 1774, he published his own work, *Von den äusserlichen Kennzichen der Fossilien* (*On the external characters of minerals*).

Werner sharply distinguished the realms of the living and the nonliving; methods developed for the former, in his view, could not be legitimately applied to the latter. Contrary to the mainstream of eighteenth-century thought, Werner denied that there was continuous gradation between the three kingdoms of nature. He emphasized the differences, rather than the analogies, between the plant and animal kingdoms on the one hand and the mineral kingdom on the other. Mineralogists, he thought, had been overly impressed by botanical and zoological taxonomy and far too uncritical in extending it to the mineral kingdom.

The difference between living and nonliving objects lay in the way they were constituted. Individual plants and animals were collections of distinct parts; the relationship between those parts provided the basis for their classification. Individual minerals were not collections of distinct parts. As Werner (1774/1962, xxvi–xxvii) explains:

> We immediately find a basic difference among [natural bodies] because they are divided into two principal kinds: in the first one, the essential features occur in their mode of association, whereas in the second, they occur in their composition. To the former belong the animals and plants; to the latter, the bodies of the mineral and meteoric kingdoms. Both, being natural bodies, are aggregated and their parts chemically associated; but the former consist of parts different from each other, which we call organs [*composita*], and in these bodies the essential features are represented by the manner in which these organs are associated to each other. The latter, on the contrary, are quite simple or consist of similar parts [*aggregata*], and their features do not correspond to the manner in which these parts are aggregated.

Put another way, if a botanist were to divide a number of plants into the smallest possible parts, he would be unable to distinguish which parts belonged to each plant. The essence of a plant, then, lay in the association of components. If a mineralogist were to divide a number of minerals into the smallest possible parts, each would still be recognizable. The essence of a mineral, then, lay in its composition.

Nor could the taxonomist get away with the claim that the living realm merged imperceptibly with the nonliving. A consideration of the "transition of natural bodies one into another," showed that "no solid argument can nor ever will be produced, . . . to prove the transition from the animal and vegetal kingdom into the mineral kingdom." Werner (1774/1962, xxviii) reiterated his reason: "The natural succession of essential features in the former follows their mode of association [*compositionibus*] and in the latter, their composition [*mixtionibus*]."

The nature of mineral composition settled the question of the essential character of minerals. "The essential features of minerals, therefore, must necessarily exist in their composition, because they cease by a change of the latter" (1774/1962, xxvii). This meant that minerals must be classified according to their composition.

> When we want to classify natural bodies, or to determine their natural arrangement, which is the same thing, we must seek some principle on which this classification can be based. This principle must exist in the nature of these bodies, because their order or arrangement is to be natural. [1774/1962, xxvi]

Since chemical composition was the essential character of minerals, they "*should be classified and the species separated on the basis of their composition*, for a mineral system has no other purpose than to determine the natural order or classification of the different minerals" (1774/1962, xxvi).

Werner followed Cronstedt in thinking that the Linnaean hierarchy, particularly its upper divisions, was inappropriate to mineral taxonomy. "Mineralogists have hitherto tried too hard to apply to their systems the four graduated divisions of logicians—classes, orders, genera, and species—thus doing in some measure violence to Nature" (1774/1962, xxix). Werner did not expound his classification in his 1774 work on the external characters of minerals. At that time he accepted Cronstedt's system. In 1789, an account of his mineral system appeared, and finally, in 1817, his *Mineral-System*, based on the distinction of the earths, salts, metals, and bituminous substances, was published.

IDENTIFICATION VERSUS CLASSIFICATION

The preceding description of Werner as an ardent advocate of the classification of minerals by composition may sound strange to readers who accept the widely held view that Werner was the main spokesman for the classification of minerals by external characters (Greene and Burke 1978). But Werner, like most other taxonomists, drew a sharp line between identification and classification. Identification is the development of a repertoire of techniques for recognizing further examples of an individual that has already been described and named. Classification is the process of assigning the entities in the world to a place in a conceptual network. They arc distinct, if related, activities. Werner thought that minerals should be *identified* by their external characters; he thought they should be *classified* by their chemical characters.

Identification takes place in two stages. First, the mineralogist describes the characters of a specimen that he assumes to be a natural kind. Second, he establishes ways to recognize other specimens of this mineral—that is, other specimens with the same cluster of properties—normally by the use of an identification key. Such keys are familiar to most people through the use of handbooks for field naturalists. Using one diagnostic character, the mineral specimen is assigned to one of two or more mutually exclusive and exhaustive groups; then the procedure is repeated until the mineral has been uniquely identified. The groups in the identification keys are not necessarily, or even usually, coincident with the classes and lower subdivisions of classifications.

Classification, Werner (1774/1962, xxv) insisted, was different from identification: "*To classify minerals in a system and to identify minerals from their exterior* are two distinct things, and . . . to attain both of them, very different means should be employed." For practical reasons, he stated that he would rather "have a mineral ill classified and well described, than well classified and ill described" (1774/1962, xxix). The problem was that mineralogists lacked a well-defined, descriptive language for identification. Minerals were more difficult to identify than animals and plants. At the end of the preceding century, Woodward (1695, 170) had complained,

> To write of metals and minerals intelligibly and with tolerable perspicuity, is a task much more difficult than to write of either Animals or Vegetables. For these carry along with them such plain and evident Notes and Characters either of Disagreement, or of Affinity with one Another, that the several Kinds of them, and the subordinate Species of each, are easily known and distinguish'd, even at first sight; the Eye alone being fully capable of judging and determining their mutual Relations, as well as their Differences.

Werner intended his essay *On the external characters of minerals* as an aid to identification, not as a treatise on classification. He was convinced that external characters offered the most convenient way to identify minerals, particularly in the mine and the field. He defined descriptive vocabularies for color, cohesion (the most important category, including solidity and crystal form), unctuosity, coldness, weight, smell, and taste. The essay so impressed Werner's contemporaries that he was offered a teaching position at Freiberg, where he remained for the rest of his life.

Others thought that external characters were too ambiguous for accurate identification. Bergman (1783, 7–8), for example, admitted that external characters offered a shortcut to identification, but argued that they were "insufficient" since "they cannot even enable us to distinguish the calcareous from other earths" (see chapter 3). Axel Cronstedt found that many minerals, when heated with a blowpipe flame, revealed characteristic color changes and melting patterns, a technique still used in the twentieth century (Oldroyd 1974b). In practice, most mineralogists used a mixture of different techniques. They were able to identify an increasing range of minerals, an ability that was reflected in the construction of increasingly precise causal theories (see chapters 5 and 8).

THE TAXONOMY OF ROCKS AND FOSSILS

Minerals were not the only objects with which mineralogists dealt; there were also rocks and fossils. For different reasons, neither of these received the systematic treatment accorded to minerals. Rocks were treated as no more than extended masses of one or more minerals for most of the century; hence, they seemed to require no special taxonomic apparatus (Bergman 1783). Toward the end of the century, classifications of rocks according to their age began to appear (see chapter 5). But it was not until the nineteenth century that petrology and stratigraphy—specialist subdisciplines devoted to the study of rocks—emerged (see chapter 7).

The case of fossils is a little more complicated, particularly as our ignorance of eighteenth-century paleontology means that my claims must be tentative. Many mineralogists and cosmogonists mentioned fossils. The majority thought they were the remains of organic beings (even though disputes continued). Furthermore, many mineralogists, following Leibniz, thought that the presence of fossils on mountaintops showed that the ocean had once covered much of what is now dry land. Nonetheless, natural historians, whether mineralogists, botanists or zoologists, appear to have done little to classify fossils, or even to offer guides to their identification.

Most mineralogists gave fossils short shrift. They included fossils in their classifications only because fossils were composed of mineral substances. After all, if fossils were the remains of organic beings, then they were as much the business of botanists or zoologists as they were the business of mineralogists. Since most natural historians did not believe that species had gone extinct, mineralogists could wait for botanists or zoologists to find and classify the living equivalents of fossil plants and animals. Linnaeus, for example, devoted only a tiny section of his mineral taxonomy to fossils, which he grouped according to their organic, not their mineral, characters. Bergman (1783, 236), although he thought fossils were "highly useful" for reconstructing past conditions, also treated them as outside the realm of mineral taxonomy.

But the botanists and zoologists thought that describing, identifying, and classifying fossils promised few rewards at a time when there were so many living species awaiting examination. They had no reason to believe that fossil species were extinct. Indeed, in view of the new species constantly being discovered around the world, they had every reason to suspect that living representatives of most fossil species would soon come to light. Under these circumstances, wasting time on the fossilized varieties seemed a folly. Generally, few specimens of fossil species were available. Those few were frequently poorly preserved, and, even when well preserved, they very rarely showed the softer parts of the plants or animals. The sensible botanist or zoologist ignored fossilized species, and concentrated on living ones. In short, well into the nineteenth century, neither mineralogists nor botanists nor zoologists put fossil taxonomy high on their list of priorities. Consequently, few fossil taxonomies seem to have been constructed during the eighteenth century. One exception was that proposed by Johann Walch (1725–1778). Walch's real interests lay in philology and mineralogy, despite his position as professor of philosophy and poetry at Jena. He came into possession of a set of copperplate engravings of fossils, prepared by a certain Georg Knorr (1705–1761) as a commercial publishing venture. With the permission of Knorr's heirs, Walch published the remainder and tried to introduce some order into the accompanying text, which appeared between 1762 and 1773. But it was not until his pupil Johann Blumenbach had argued that extinction occurred, and mineralogists had independently decided that fossils were excellent indicators of the age of rocks, that paleontological systematics began to flourish (see chapter 7).

CLASSIFICATION AND THE STAGES OF SCIENTIFIC DEVELOPMENT

Two theses, both of them with many adherents, link taxonomy to specific stages within the development of science. The first thesis is that taxonomy is characteristic of early "immature" science; the second thesis, put forward by Foucault, is that the type of taxonomy advocated by Linnaeus was typical of the European "classical age" from 1650 to about 1800. Since both of these theses seem to me to be fatally flawed, I shall conclude this chapter by indicating why neither thesis is supported by eighteenth-century mineral taxonomy.

The link between taxonomy and immaturity in science is widely accepted by philosophers. Kuhn (1962), for example, relegates seventeenth-century natural history to pre-paradigmatic, immature science, describing it as a "morass." Hesse, who is more sympathetic to problems of classification than most philosophers, uses the immaturity thesis to explain why so few philosophers discuss taxonomy. Philosophers, she says, have concentrated on sciences like physics, in which problems of causality have replaced problems of classification; "physics and chemistry passed the classification stage long ago" (Hesse 1974, 73). As a result, although new laws are still proposed, "new predicates rarely have to be introduced into the language to refer to newly significant resemblances."

As a claim about physics, it is dubious enough. At a time when physicists are taking positive delight in dreaming up names like *color*, *charm*, and *spin* for new predicates, the evidence for the immaturity thesis seems extremely slim. But worse, it misses the point that philosophers of biology have recognized—namely, that the contrast between causal theories and classification is spurious. Since scientists—mineralogists and geologists included—frame theories to account for causal interactions between entities, they have to satisfy themselves that they have correctly specified not only the causes but also the entities. Since we have enormous historical evidence that such specifications have been fallible, we should not expect taxonomic questions to be settled once and for all in the early stages of a particular science's development. On the contrary, we should expect problems about classification to recur throughout the development of the science, as is now the case in physics.

Foucault (1966/1973) places classical taxonomic theory differently; he does not see it as a stage occurring in the development of every science, but as typical of one particular period, the "classical age" in European thought. Classifications, according to him, are also to be found in the preceding and succeeding periods. In the preceding period, classification was based on "similitude"; in the succeeding period, on "succession." In the classical period, it was based on "representation"—that is, on the exact description of external, visible structures. The belief that the "essences" of bodies were faithfully "represented" by their external features, and that these features themselves were subject to exact analysis, made possible the establishment of ordered, hierarchical tables of classification based on the systematic comparison of visible structures. The classical age, he claims, was bounded by sharp ruptures around 1660 and 1800.

Foucault restricts his analysis to zoology, botany, economics, and linguistics. It seems reasonable to believe that he would also have expected it to apply to mineralogy. In an interesting article, Albury and Oldroyd (1977) have explored this possibility. They conclude that it works well for the early part of the period, if chemical mineralogy is excepted, less well for the later part of the period as sharper distinctions were drawn between the living and the nonliving. I, however, am less sanguine than Albury and Oldroyd about the fruitfulness of this line of analysis. It is not that Foucault has completely misrepresented classification in the period, but that his scheme is so rigid as to hinder rather than help historical understanding. At least in mineralogy, I do not detect sharp ruptures, certainly not at Foucault's proposed dates. Mineralogists from Aristotle to Agricola had classified minerals in part by their external, visible features. Some—Haüy is an example—continued to do so into the nineteenth century, well past the date of the second purported rupture. If the evidence for Foucault's ruptures is slim, so, too, is the evidence for the exclusive use of "representation" during the classical period. Chemical mineralogy continued to flourish throughout the eighteenth century, arguably being the majority position. In short, the similarities between different fields of inquiry and the changes in theoretical orientation over time can be explained more simply by the historians' time-honored concept of intellectual traditions than by resorting to vague Foucaultian theses about "the positive unconscious of knowledge" and "ruptures in human consciousness" (Foucault 1966/1973, xi).

5
Werner and the
"School of Freiberg"

In 1786, Werner published his *Short Classification and Description of the Various Rocks* (Ospovat 1967, 1969, 1971). In this, in his *New Theory of the Formation of Veins* (1791), and in his lectures, he propagated a theory of the earth that dominated geology until the late 1820s and that contained the seeds of the separation of causal geology from historical geology (Wagenbreth 1955, 1967, 1980; Guntau and Mühlfriedel 1968). Hence, Werner's theory marks a pivotal point in the development of geology. The succeeding generation looked to Werner and to his "school of Freiberg" as the source of most of their geological theories.

Werner was steeped in the tradition of chemical cosmogony. He traced his heritage all the way back through Oppel, Bergman, Wallerius, Lehmann, Hoffmann, and Henckel to Becher, Stahl, and Agricola (Werner 1791/1809). Unlike many of his predecessors, Werner was not himself a chemist. By the time he took a teaching position at Freiberg, the school was large enough that he could concentrate on mineralogy and geognosy, leaving chemistry to others. Nonetheless, he took a keen interest in chemistry. He had studied it as a student at Freiberg, and later as a teacher there he regularly interacted with chemists. He helped organize the construction of a new laboratory for the Academy in 1797 and supported the appointment of W. A. Lampadius to teach the new chemistry of Lavoisier (Ospovat 1971). He was indebted to the chemical cosmogonists for the theory of the different earths, for the theory of the successive deposition of rocks from aqueous solution, and for the definition of rocks in terms of age as well as mineralogy. His writings make clear that he recognized his debt. Werner (1791/1809, 101) insisted that, in order to appreciate his work, his pupils needed

to have a just conception of the difference between a chemical precipitate and a mechanical deposition; to have a perfect idea of

simple elementary bodies which are not susceptible of any transmutation; to be acquainted with the theory of solution and precipitation founded upon chemical affinities; and above all, to know
that the same individual solution may furnish, not only at the same
time, but also in succession, precipitates of a different nature.

But during the course of his career, Werner transformed the
Becher-Stahl tradition from which he had taken so much. He made
the time of formation of rocks, not their mineralogy, their most
important character. Well aware that he was flouting the precepts of
taxonomy, he named bodies of rock formed in the same period
"formations," and made these historical entities—formations—
more important than chemical ones. He concentrated on the earths
at the expense of the metals, and on rocks at the expense of veins.

Werner's adoption of the term *geognosy* signaled this change of
emphasis. He gave it a much wider definition than mineralogy, a
definition that closely corresponds to the nineteenth-century use of
geology. For Werner, "geognosy is that part of mineralogy which
acquaints us systematically and thoroughly with the solid earth, that
is, with its relationship to those natural bodies that surround it and
which are familiar to us, and also, especially, with the circumstances
of its external and internal formation and the minerals of which it
consists according to their differences and mode of formation." He
chose the term because "*gnosis* means the abstract systematic
knowledge of something, and . . . geognosy taken verbatim can be
called the abstract systematic knowledge of the solid earth" (Ospovat 1971, 102).

To demonstrate Werner's debt to the past, I shall begin by
describing his overall system, reconstructing it from the writings of
his pupils as well as from his elliptical remarks in published works.
I shall then describe his major innovation—the concept of a "formation"—and trace its consequences for the study of the earth.
Finally I shall show how the influence of Werner's school of
Freiberg went beyond Saxony, to all Germany, to continental
Europe, and to all other parts of the world in which geologists
worked and traveled.

WERNER'S ACCOUNT OF EARTH HISTORY

Werner's *Short Classification* includes only a fraction of the materials that Werner covered in his course of geognosy. He began by
setting geognosy in the context of mineralogy and the other sciences, by outlining its history, methods, and uses. He then de-

scribed the earth's place in the solar system and discussed its origin. He talked about the figure of the earth, the causal processes, particularly aqueous ones, affecting its face, the formation of rocks, and the role of fossils in interpreting the past of the earth. Only then did he get down to the classification and description of the rocks (Ospovat 1967).

Werner accepted the ruling presumption that the earth had once been fluid. Like his predecessors in the chemical tradition, he believed that a preliminary, but incomplete, sorting of earth, water, air, and fire into concentric layers according to their weight preceded the period in which chemical processes predominated (Ospovat 1979). But Werner was impatient with speculation about the earth's distant past and deep interior:

> We want to leave the dark abyss of the earth, where we can only wander about along the giddy paths of conjecture and return to that upper region of the solid earth, where the light of experience of searching minds illuminates the path. [quoted by Ospovat 1979, from "Notes on Geognosie," Werner manuscript]

In this upper region, the rocks provided testimony about the earth's past.

Again following his predecessors, Werner believed that the rocks had been deposited from water. Like Henckel, Wallerius, and others, he believed that the original ocean was a thick mixture of minerals and water, quite unlike the thin fluid with dissolved salts that forms the present ocean. The only way to determine the former contents of the ocean was to examine the rocks that it had deposited. Werner quite probably thought that the original ocean had also been hotter, since, as we have seen, "siliceous earth" (silica) was known to be soluble in hot, basic solutions, but only minimally soluble in cool ones.

Following well-established precedent, Werner called the rocks lowest in the succession (and hence presumably the oldest)—the "primitive class." They were composed of relatively insoluble "siliceous [quartz and sandstone], argillaceous [clay and shale] and magnesian earths," and metals and their ores. But they included almost none of the more soluble earths, or the salts or bitumens (Jameson [1808] 1976). Many of the rocks were crystalline, suggesting that they had formed in still, presumably deep water. Their uneven surface, ranging from low ground to the highest mountaintops, supported this conclusion. Werner believed that the primitive rocks had crystallized on the uneven surface of the primeval earth from an ocean deep enough to cover the highest mountains. He divided them into a regular and determinate succession of individual formations, in order: granite, gneiss, mica-slate, clay-slate, porphyry-slate, porphyry, basalt (in the first version of his classification), amygdaloid, serpentine, primitive limestone, quartz, and topaz-rock. Above the primitive rocks, Werner subsequently inserted a small class to which he gave the self-explanatory title of the "transition class."

Thus far this was all quite conventional. The observed order of the rocks and the order predicted by chemical theory seemed to agree. But Werner made a further claim that suggests he gave historical evidence precedence over chemical analogy; he argued that mechanical precipitates in the succession showed that the ocean had retreated at one point, only to readvance, though not to its former level (table 2).

Werner believed that the composition of the ocean changed gradually over time, rather than in sudden jumps. Hence, he expected to find gradations from one formation to the next. For him, though, as we shall see, not for his successors, the determinability of the formations did not mean that they were separated by sharp breaks. Werner was no "catastrophist." He believed that the processes going on today were the same as those in the past. Observations in the field demonstrated that

> several mechanical and chemical depositions . . . are daily taking place, as it were, under the eye; . . . a comparison of their structure, with that of the great fossil masses of which the earth is composed, evinces so complete an agreement, as entitles us to infer with great certainty, that these also have been formed by the same agent. [Jameson 1808/1976, 75]

As we shall see in chapter 8, the Wernerians made extensive investigations into the kind and intensity of contemporary causal processes.

Werner's next class, the layered ("*Flötz*") or stratified rocks, again corresponded to a group identified by his predecessors (Werner 1786/1971). Some *Flötz* were crystalline; others were "concreted" or "cemented." Consequently, Werner concluded that "mechanical" as well as "chemical" deposition had taken place (Jameson 1808). During much of this period, the ocean must have been more turbulent. This was consistent with the lower elevation of the stratified rocks, which indicated that the ocean had retreated after the deposition of the primitive class. Once again, Werner paid

Table 2. Werner's Theory of the Stages of Earth History

Predominant Lithology	Granite, gneiss, slate, limestone, porphyry, serpentine		Greywacke, transition limestone		Sandstone, limestone, coal, gypsum, basal-clay		Alluvial, Volcanic
Structure of Rocks	Crystalline	Less crystalline	Still less crystalline	Stony	Stony	Stony and crystalline	
Record of Life					Organic fossils occur in increasing complexity		
Position of Rocks	Form the highest mountains and underlie all other rocks		Sloping up against primitive mountains		Predominantly lowlying, almost horizontal strata		
Name of Class	PRIMITIVE		TRANSITION		FLÖTZ		RECENT
Postulated Turbulence of Water	Calm	Less calm	Less calm again	Stormy	Stormy	Stormy Calm	
Postulated Height of Water							

Based on Ospovat's introduction to Werner's *Short Classification*, 1971, 22. Werner's theory is intermediate between genetic theories of earth history in terms of changing chemical composition of the ocean and strictly historical reconstructions based entirely on the testimony of the rocks. The size of the various time periods is arbitrarily drawn and should not be taken to represent Werner's absolute time scale.

careful attention to detail. He argued that there must have been two inundations during the *Flötz*, separated by periods when the ocean stood at lower levels.

Readily soluble earths, such as the calcareous and gypsous, appeared in quantity in the *Flötz*. So did rock salt and bituminous substances such as coal. Most of these rocks contained organic fossils; and the higher in the succession, the greater their variety and complexity.

> When the water diminished in height, and the dry land began to appear, marine plants, and the lowest and most imperfect animals, were created. As the water diminished, it appears to have become gradually more fitted for the support of animals and vegetables, as we find them increasing in number, variety and perfection, and approaching more to the nature of those in the present seas. [Jameson 1808/1976, 82]

Werner was less interested in the two top classes, which he believed to be of recent origin. The alluvial class was composed of isolated mechanical deposits of sand, clay, loam, and coal. Present-day agents, particularly water in the form of rain, rivers, and the ocean, were wearing away the land in many places. The detritus was deposited in valleys and on the ocean floor, but these formations were much more restricted in extent than the older formations. Werner suggested these rocks had been formed by the destruction of previously deposited primitive and *Flötz* rocks.

In the volcanic class, Werner included only those rocks, such as pumice and lava, that were still being produced by volcanoes. Since Werner, like many others, thought that volcanoes were caused by burning beds of coal, and since coal was not found below the stratified rocks, he believed volcanic products to be late and anomalous. Early in his career, Werner had been prepared to consider the possibility that basalt was of igneous origin. Basalt was a dark claylike rock that some mineralogists had decided was volcanic on the basis of field evidence (see chapter 8). However, a visit to the famous basalt outcrop at Stolpen in 1776 convinced him that it looked just like other layered rocks, and by the time he published the *Short Classification* he had decided that all basalts were aqueous in origin. The field evidence did not decide the matter, at least not in Saxony. R. E. Raspe thought that the Saxon basalts were igneous in origin; J. F. W. von Charpentier, Werner's colleague and an ardent vulcanist, agreed with Werner that they were not (Raspe 1776; Charpentier 1778; see also Carozzi 1969; Ospovat

1980). In chapter 8, we shall pursue this question further. But for now suffice it to say that there was no clear consensus on the matter.

Finally, again following his predecessors, Werner (1791/1809) suggested that veins were formed in "rents." These rents had appeared as the wet beds dried, shrank, and cracked, partly under their own weight, partly as a result of dessication. Subsequently, they were filled by minerals deposited when the ocean advanced again.

Werner did not know what had caused the ocean's retreat, nor why there had been occasional reversals in this retreat. Helmont's theory that water (presumably including oceanic water) could be transmuted into earth had been thoroughly discredited, despite a belated rescue effort by Wallerius. Werner also rejected the theory, common among his predecessors in the Becher-Stahl tradition, that the water had retreated into the abyss or into caverns in the earth. Nor did the suggestion that it had been drawn away by astronomical causes appeal.

But Werner was not overly concerned by his inability to identify the cause. Given the strength of the evidence, argued Werner (1786/1971), the existence of the phenomenon should be accepted even in the absence of an explanation of its cause. Jameson (1808/1976, 6:82–83) agreed: "We may be fully convinced of . . . [the] truth [of the theory of the diminution of the waters], and are so, although we may not be able to explain it. To know from observation that a great phenomenon took place, is a very different thing from ascertaining how it happened."

Nevertheless, when challenged, Werner had a couple of responses ready. Sometimes he pointed out that the problem was a figment of our peculiar human viewpoint. The depth of the ocean was nothing compared with the overall dimensions of the globe. An ocean that had covered the highest mountains might seem enormous by human standards, but by global standards it was the thinnest of layers, so that the disappearance of the ocean was largely a pseudoproblem (Ospovat 1979). Later, in 1813, Werner invoked Lavoisier's new chemistry. Far from being an elementary substance, as chemists had taught when Werner was a student, Lavoisier had shown water to be a combination of the "airs," hydrogen and oxygen. Consequently, Werner had "no doubt that nature used a large part of the universal ocean to form the atmosphere" (Ospovat 1971, 23). In suggesting this, Werner was reinterpreting older ideas. Wallerius had suggested that water might be transmuted into air in order to explain why Genesis failed to mention the formation of the atmosphere (Oldroyd 1974).

Werner thought it unlikely that original ocean had contained all the materials that had subsequently been deposited to form the rocks. The presence of rocks of the same mineralogy at different points in the succession ("formation suites") suggested to him that new chemicals had been added to the ocean at different periods. Werner did not say what he thought had caused this, though we may speculate that he had the erosion of earlier formations in mind.

Thus far the system I have been describing falls within the tradition of chemical cosmogony, modified only by changes necessitated by increasing field knowledge and by developments in cognate disciplines, particularly chemistry. Since Werner was continuing and extending the chemical tradition, his contemporaries found the theory both familiar and plausible. Subsequently the arguments for Werner's theory were forgotten, so that Gillispie (1951/1959, 45) could claim: "Objections to [Werner's] conception are so obvious that it is difficult to understand how responsible geologists could ever have accepted it."

But when geologists referred to Werner, they did not single out his general theory of the earth for attention but his concept of "formations" and his assertion of their invariable order world-wide.

FORMATIONS: TIME IS OF THE ESSENCE

Werner insisted that the "essential differences" between rocks of various kinds were "mode and time of formation" (Werner manuscript, quoted by Ospovat 1971, 19). In line with this definition, he coined the term *Gebirgsformation,* or "rock formation," to describe the major rocks making up the earth's surface (Ospovat 1971). This was usually translated into English as "formation" and is the term I shall henceforth use, since it signals a very important shift in the development of geology.

Werner's phrasing of his definition is crucial. It indicates that he believed that the geologist should group together rocks of the same age and mode of formation, even if their other characteristics, such as mineral composition, varied. Thus, Werner rejected the two standard ways of distinguishing rocks; he followed neither the miners' use of method of working, extent, and location, nor the mineralogists' use of constituent minerals. He was not the very first to do this, but was anticipated by Füchsel, Lehmann, and others. Nonetheless Werner was the one who made the formation the central concept of historical geology.

As we have seen, Werner was well aware of accepted taxonomic practice. When he chose mode and time of formation as the essential characters of rocks, he knew what he was doing. And that was standing taxonomic practice on its head. As we have seen, the purpose of taxonomy was to sort out natural kinds—that is, entities that, among other things, were not temporally restricted. By making temporal restriction a defining characteristic of formations, by making time of the essence, Werner defined formations as unique, historical entities, not as natural kinds.

Werner may well have believed that mode of formation and time of formation co-varied. Certainly most cosmogonists within the Becher-Stahl tradition would have thought so. Werner does not seem to have addressed the matter directly, but his successors decided that this was not the case. Consequently, geologists dropped mode of formation (which had always been the less important character) and based their classifications entirely on age (see chapter 7).

Such a classification (and some, such as Humboldt, even doubted that it could be called a classification) was intrinsically historical. It was totally at odds with the static view of the world embodied in earlier natural histories. It also distanced geology from natural philosophy and chemistry, for the things that the physicist or chemist had to say about formations now became of only secondary interest to the geologist. Thus historical geology drew away from the disciplines that had informed mineralogy and chemical cosmogony.

The classification of rocks by age was a simple linear structure compared with the maplike structures of the mineralogists or the branching hierarchies of the botanists. Each formation was composed of a sequence of beds, and sequences of formations made up the primitive, transition, secondary, alluvial, and volcanic classes. The bulk of the earth was made up of "universal formations" (Jameson 1808/1976). Unlike "partial or anomalous" formations, they extended around the whole globe. Their mineral content varied from place to place, reflecting the uneven composition of the ocean from which they had been deposited. Since that time, most of them had been at least partially destroyed by erosion, so that they were no longer continuous. These universal formations, far from being unlimited in number and impossible to order, were "very distinct and determinable" (Werner 1786/1971, 38). They numbered between twenty and thirty, Werner estimated, more than the three or four groups previously distinguished by mineralogists. They were also far fewer than the fine divisions in mining logs. Late-eighteenth- and early-nineteenth-century mining records demonstrate how hard it was to generalize from the uncorrelated lists of dozens of different strata, often only a few feet or a few inches in thickness. Werner's clear description of an intermediate number was a crucial step that brought order to a previously chaotic area of enquiry. Werner's list was short enough to be applied and long enough to encompass many of the different formations on the earth.

But deciding to arrange the rocks by their age was one thing; doing it was quite another. Age (or mode of formation, for that matter) cannot be observed, at least not directly. The form of the classification was simple; filling in its details was not. By redefining the essence of a rock as its age (which was not observable) instead of its mineralogy (which was), Werner introduced a new problem for the geologist—namely, that of constructing some bridge between a formation's observable characters and its age.

The solution was ready at hand in the tradition from which Werner came, though it had not been addressed so explicitly before. If, as those in the Becher-Stahl tradition assumed, all the earth's rocks had been sequentially laid down in water, and if they had not been subsequently disturbed, then position in the succession was an unambiguous indicator of age. The upper formations in a sequence had been deposited after the underlying ones, and were thus younger. This principle of superposition was not limited to the chemical tradition. As we have seen, Steno, who was highly regarded by Werner, had clearly articulated it for stratified rocks (Ospovat 1980).

I may seem to be belaboring this point, for the "principle of superposition" is one of the first things that an undergraduate learns in a geology class. But it is important to stress that it was an assumption, and an assumption that not all geologists believed to be reliable. Those cosmogonists who believed that the earth had cooled from the outside in, or who asserted that the lower rocks had been intruded into the upper rocks, could not use the order of superposition as a universal indicator of time of formation. Rather, it was an indicator restricted to the small number of strata that they conceded had been laid down in water. By contrast, Werner (1791/1809, 3) confidently asserted that the observable character of position in the rock succession was a reliable indicator of the essential, but unobservable, character of time of formation.

His confidence turned out to be somewhat premature. In the succeeding generation the recognition of the igneous origin of a wide range of rocks, the recognition of metamorphism, and the recognition of the scale of displacement to which strata were subject once they had been deposited made the sorting out of the order of succession much more difficult than Werner had anticipated. Nonetheless, the principle of superposition offered a key to at least a preliminary sorting of the formations. With this as a base, historical geologists were able to integrate trickier cases into the succession.

But classification was not dear to Werner's heart; indeed, this may be the reason he was so cavalier about making a temporally restricted entity like the formation the basis of his geological classification. For Werner, identification (or "description," as he called it in his title *A Short Classification and Description of the Various Rocks*) was the first order of business, in geology as in mineralogy. Possibly it was even more important in geology than in mineralogy. Since the entire vertical sequence of formations cannot be observed in a single exposure, geologists had to assemble the succession by correlating limited sections from exposures that were often miles, or even continents, apart. This meant that reliable ways of identifying rocks had to be found. To facilitate identification and correlation, Werner described the observable characters of each of the formations he distinguished. In the *Short Classification*, published some thirty years before his death, he gave pride of place to mineralogy and secondary status to relief, dip, texture, and fossil content.

There is every reason to believe that, over the years, Werner modified this order of priority. And if not Werner, then his pupils certainly modified it (see chapter 7). Mineralogy declined in importance as geologists, including Werner, realized that it did not indicate age reliably. Indeed, Werner had suspected this from the time that the *Short Classification* first appeared. His "formation suites" were sequences of rocks of a given mineralogy that occurred more than once in the succession (table 3). Limestone, for example, appeared not just once but several times in the succession, although often in a modified form. Similarly, slate, trap, porphyry, gypsum, salt, coal, and serpentine all recurred on a regular basis.

At first, Werner and his followers tried to relate these modifications to age, so as not to confound the bases of his classification. Jameson ([1808] 1976, 87) stated, "All the members [of such a series] have general characters of agreement, and the individual members bear characters expressive, not only of the period of their formation, but also of the circumstances under which they were formed."

The differences in rock texture between the members of a formation suite indicated the shifting composition of the ocean at different periods, and hence the different circumstances under which

Table 3. Werner's System of Formation Suites

FORMATION

PERIOD.	SLATE.	LIMESTONE.	TRAP.
PRIMITIVE *Stratification conformable*, with finking levels of the out goings of the newer and newer ftrata.	Granite. Gneifs. Mica flate. Clay flate.	Limeftone. Limeftone. Limeftone.	{ Hornblende-flate. Hornblende-rock. Primitive gre nitone. Greenftone flate
Stratification overlying.	Neweft Granite.		
TRANSITION	Grey-wacke. Grey-wacke-flate.	Tranfition-Limeftone.	Greenftone. Amygdaloid.
FLŒTZ. *Stratification conformable*, with linking levels of the outgoings of the newer and newer ftrata.	1ft Flœtz Sandftone. 2d Flœtz-Sandftone. 3d Flœtz Sand. ftone. Sandftone in the coal formation	1ft Flœtz Lime-ftone. 2d Flœtz-Lime ftone. Chalk. Limeftone and Marl in the Coal formation.	Amygdaloid. Greenftone and Amygdaloid in the Coal-formation.
Stratification overlying.	Quartzy-fand-ftone, in the Neweft Flœtz-trap formation.	Limeftone in the Neweft Flœtz trap formation.	Greenftone. *Grayftone. Porphyry-flate. Bafalt. Wacke Trap-Tuff.* Amygdaloid. Iron clay.
ALLUVIAL.	Clay. Loam. Sand Gravel. Rolled maffes.	Calc-fpar. Calc-tuff. Calc-finter.	

Table 3 (Continued)

SUITES.

PORPHYRY	GYPSUM.	COAL.	SERPENTINE.
Oldeſt Porphyry. Oldeſt Porphyry. Oldeſt Porphyry.	} - - Uldeſt Gypſum. } - -	⎰ Slaty glance-coal ⎱ Slaty glance-coal Slaty glance-coal Alſo ⎩ Graphite.	Oldeſt Serpentine. Oldeſt Serpentine. Oldeſt Serpentine.
Newer Porphyry, including Ilay-P rphyry, Pitchſtone-P. Obſidian P. Pearlſtone P. -Alſo S enite.			Newer Serpentine.
	Tranſition- Gypſum.	Slaty glance-coal. Graphite ?	
Porphyritic-Stone in the independ- ent Coal-forma- tion. It may be conſidered as the 3d Porphyry-for- mation.	ıſt Flœtz-Gypſum. Rock Salt Forma- tion. 2d Flœtz Gypſum.	ın the indepen- ent Coal forma- tion : Coarſe-Coal. Foliated-Coal. Slate-Coal. Cannel Coal. Pitch-Coal. Slatyglance coal. Graphite.	
Clay-ſtone ? Compact felſpar ? Pitchſtone-Por- phyry. Obſidian-Porphy- ry Pearlſtone-Por- phyry.		In the Flœtz-trap formation : Pitch-Coal. Columnar-Coal. Conchoidal glance- coal.—alſo Bituminous Wood. Alum-earth. Common Brown- Coal. Moor-Coal	
		Earth-Coal. bituminous-Wood Alſo, Moor-Coal. Common Brown- coal.—And Alum earth.	

From Jameson 1804–8, 3: 96–97. The Wernerians replaced the older genetic theory that particular rocks were laid down in particular time periods with a more sophisticated account of the changing character of different rock types over time.

the rocks were formed. For example, limestones in the primitive period were pure white and crystalline, in the *Flötz* they were granular and marly, and in recent caves and springs they were incrustations. But by the early nineteenth century, most Wernerians acknowledged that rocks of essentially identical mineralogy could occur several times in the succession, necessitating the search for other indicators of relative age (see chapter 7).

As mineralogy waned in importance as the identifying character of formations, fossil content waxed. Werner had always been interested in fossils (Wagenbreth 1968). He had a respectable collection of fossils that his students used (Ospovat 1971). Like many of his predecessors, he thought fossils were one of the characters that set the secondary rocks off from the primitive ones, and he knew that these became more various and complex higher in the geological column. At a general level, he was aware that different formations contained different fossils, though he did not identify these as precisely as his successors were to do. From 1799, he offered a course in paleontology (Ospovat 1971).

Many of Werner's students pursued paleontology with his encouragement. Humboldt (1823, 58) described how:

> Although [Werner] did not possess the necessary means for obtaining a rigorous determination of fossil species, he never failed, in his course of lectures, to fix the attention of his pupils on the relations that exist between certain fossils and formations of different ages. I witnessed the high satisfaction which he felt, when M. de Schlottheim [sic], one of the most distinguished geognosts of the school of Freiberg, began in 1792 to make these relations the principal object of his studies.

Schlotheim (1765–1832), Karl von Hoff (1771–1837), and Alexander von Humboldt (1769–1859), all studied with Johann Friedrich Blumenbach (1752–1830), who was instrumental in promoting paleontology in Germany. The University of Göttingen, where Blumenbach taught, was unusually sympathetic to disinterested learning, unlike many other German institutions of higher learning, notably the mining academies, which stressed utility. Consequently, Blumenbach was free to theorize and speculate, which he did under the banner of *Naturphilosophie* (LeNoir 1981). Since he had briefly studied under Walch in Jena, he was aware of what little work had been done on fossils in Germany. In addition to many of Werner's own pupils, other figures sympathetic to Wernerian theory studied with Blumenbach. They included Karl von Leonhard (1779–1862), the leading petrographer of the early nineteenth century, and K. F. Kielmayr (1765–1844), who introduced Cuvier (1769–1832) to "philosophical natural history" (Coleman 1964).

Werner's list of the identifying characters of formations allowed the correlation of formations in different regions. Without correlation between formation in different locations, knowledge of the rocks and strata remained excessively and inaccessibly local. Individual surveys of particular regions remained as parochial as, say, Whitehurst's geological investigations of Derbyshire (even though Whitehurst [1778] included a section from Shropshire and northern Ireland). A general conceptual framework to order the growing body of data about the earth's rocks was desperately needed in the eighteenth century. Whether Werner constructed this framework by observations in Saxony or, as recent evidence suggests, his grounds were more theoretical is irrelevant (Ospovat 1980). As we have seen, Werner was heir to a long tradition of mineralogists familiar with the outlines of Saxon geology, who had found that the order of the formations predicted by chemical theory was consistent with the order observed in the field. Werner offered a way of ordering the diverse local successions that miners and mineralogists around the world described. Of course, Werner was not the first geologist to make global claims. All the cosmogonists had believed their theories to hold worldwide. Werner struck a middle ground between claims, such as those of the cosmogonists, that were so general that they could not be tested in the field, and claims, such as those of the miners, that were so specific that they could not be extended beyond the area in which they were formulated. The significance of his contribution can be gauged by the fact that British mineralogists and geologists still lacked a general structure for organizing their information about the strata (Porter 1977).

Correlation and fieldwork went hand in hand. Chemical cosmogonists, who had based their classification of rocks on mineral composition, had had no reason to spend time outdoors. Having once collected their specimens, they could identify them in the laboratory. As rocks and formations became the key geological entities in the 1770s and 1780s, fieldwork became more important and laboratory experiments less important as a source of data for historical geology. Utilitarian mineral geography gave way to historical geology. Werner substituted "formations" for rocks, time slices for lithology. Succession, not matter theory, became the organizing principle for geological entities. The abandonment of the precepts of systematic natural history reinforced geology's separation from natural history.

Hence, fieldwork became the order of the day.[1] Like most of those associated with the Saxon mining service, Werner frequently had occasion to descend into mines in the course of his work. He had traveled in Lusatia, Silesia, the Erzgebirge, Bohemia, Franconia, Hesse, and Thuringia. He regretted that his job and his limited funds prevented him from traveling more extensively (Werner 1791/1809). To make up for this, he studied reports from other parts of Europe and the rest of the world and sent his pupils out to survey regions of particular interest.

Werner gave mineral geography a rationale independent of its utilitarian origins. Mineral geography ceased to be merely a matter of making gazeteers of a kingdom's mineral resources, since a survey of the formations outcropping at the land surface, defined in terms of their ages, not their mineralogies, was only tangentially relevant to economic issues. Mineral geographers now aimed to trace the formations and to reconstruct the history of the earth, not to discover and catalog its mineral resources. Werner supervised the first comprehensive survey of Saxony, issuing precise instructions on how to proceed. The survey was finally completed after Werner's death (A. Ospovat, personal communication, 6 May 1985). Werner also laid out rules for coloring geological maps, which Jameson translated and published without acknowledging the authorship (Jameson 1811; Ospovat 1971). J. F. W. von Charpentier, a colleague and contemporary of Werner's at the Freiberg mining academy, prepared a minerological map of Saxony, using color wash to show the area of outcrop of each rock as Werner suggested (Charpentier 1778). He distinguished three groups of rock within Werner's *Greywacke*, as well as the *Flötz* rocks. Charpentier's map showing the formations (fig. 7) should be contrasted with the earlier mineralogical map of Guettard and Monnet.

THE WERNERIAN RADIATION

I have made frequent reference to Werner's pupils. I now need to explain why I identify certain geologists as "Wernerians." This raises the general issue of how to characterize units of intellectual change (and their social corollaries) at a scale larger than the individual scientist.

One possibility would be to accept Kuhn's concepts of "paradigm" and "normal science" (Kuhn 1962). On such an analysis, one would say that Werner formulated a "paradigm" that dominated geology and that Werner's followers were pursuing "normal sci-

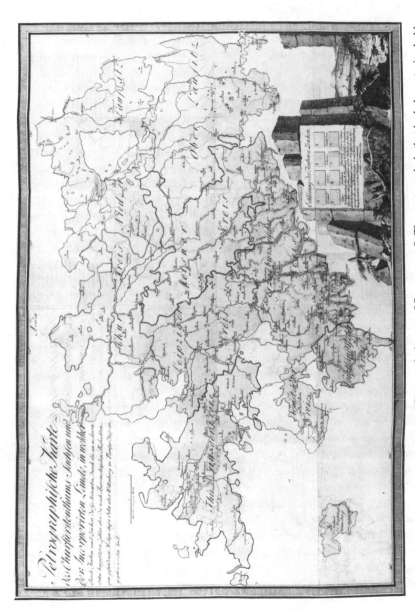

Fig. 7. A German mineral geography (Charpentier's map of Saxony, 1778. The spot symbols that had characterized older mineralogical maps are here supplemented by color washes to show the extent of formations.

ence." This approach is appealing at first glance. Werner certainly did put forward a comprehensive package of aims, methods, and theories for geologists that seems to qualify as a paradigm. Key examples were canonized in textbooks, as Kuhn suggests. Many geologists subsequent to Werner engaged in puzzle solving, unaffected by developments in other disciplines, as Kuhn claims is characteristic of normal science.

In spite of its initial appeal, however, Kuhn's theory will not do as a description of geology in the half century after Werner. A key feature of Kuhn's normal science is that the central assumptions of the paradigm are immune from criticism. As we shall see, this was simply not the case with Werner's geology. Geologists in the succeeding generation criticized and modified major parts of it, depending on their interests and background and the problems on which they worked. Unless we restrict the term *Wernerian* to that handful of disciples who shared all Werner's basic assumptions, we have to reject Kuhn's analysis. And such a restriction so narrows the scope of Werner's influence that it fails to do justice to the historical record. To refuse to call Humboldt, von Buch, or Jameson (all of whom modified Werner's ideas) Wernerian is to reduce the telling of the history of science to a series of assertions about disconnected individuals. Werner's scheme acted as a guiding framework, not as unassailable dogma. In short, there was none of the shunning of fundamental criticism that Kuhn identified as characteristic of normal science.

For similar reasons, Werner's followers cannot be said to have followed a "research programme" (Lakatos 1978). True, Lakatos's insistence on the role of a positive heuristic in science is suggestive, for Werner did give geologists guidance about how to go about extending and modifying his theory. But since the "research programme," like the "paradigm," involves an untouchable hard core of assumptions, it is equally inapplicable.

Some historians, realizing that not all geologists shared exactly the same views, reject the idea that we can talk about any coherent framework for geology in the early nineteenth century. Thus Roy Porter (1977, 216–17), has argued, "The coherence of geology [did not] depend on achieving a paradigm, in Kuhn's stronger sense of an all-embracing theory. Early nineteenth-century geology possessed no such theory—indeed it rejected the very idea."

But this, I would argue, is to put too strong a set of restrictions on what counts as commitment to a comprehensive theory. There was an over-arching theory in early nineteenth-century geology, namely

Werner's. We need a characterization of such theories that is more flexible than Kuhn's paradigm or Lakatos's "research programme."

I believe that "research traditions," which do allow changes in the basic assumptions of a theory over time, describe the reaction to Werner better than "normal science" or "research programmes" (Laudan 1977). But since I want to describe the movement to which Werner gave rise, rather than looking back at his ancestors, I shall substitute the term *radiation*, which suggests heritage and influence combined with spread and divergence. What is important is that each of Werner's successors could trace a coherent lineage back to Werner himself. Thus, to say that someone was a Wernerian is to say that a direct line of influence can be traced back to Werner by personal contact, education, reading, or any of the other ways in which one scientist learns about another's work. A Wernerian had to adopt Werner's position on some issues, though not on every issue. And he did not necessarily have to adopt the same doctrines as other Wernerians. I count someone as a Wernerian if they accept a substantial part of Werner's theory and if they are conscious that it is Werner's. The term *radiation* does not suggest that all Werner's followers adopted the same set of his claims or that they all modified the same set of claims simultaneously.

One other theoretical point needs to be made before looking at the substance of the Wernerian radiation, and that has to do with the demise of Wernerianism. Wernerianism was not replaced by another system at a single point in time by a dramatic shift, or revolution, if you will. Here, too, the Kuhnian model of normal science followed by a revolution breaks down. Instead, geologists gradually modified Werner's theory until it was transformed out of all recognition. This occurred sometime in the late 1820s. But it faded away, rather than being overthrown.

Werner's followers modified both the historical and causal aspects of his theory. Indeed, from about 1810 on, the Wernerian radiation split into two main branches, though some of the more prominent Wernerians pursued both branches. The historical branch, headed by Schlotheim, Freiesleben, von Buch, and Humboldt in Germany and Cuvier in France, accepted Werner's definition of a "formation," carried out the business of establishing the succession, surveyed new areas, and, most important, took Werner's hints about fossils and made them the main means of correlating formations and reconstructing the earth's past (see chapter 7). Although a tiny skeptical minority doubted that the concept of a formation had any utility (Greenough 1819), most geologists accepted formations as the key to historical geology.

Between 1790 and 1830, Wernerians in the historical school surveyed most of the lands then occupied by Europeans. Within Germany, von Buch and Karsten mapped Silesia, Voigt (1752–1821) surveyed the duchies of Weimar and Eisenach between 1781 and 1785, Matthias Flurl took on Bavaria; Mohs, the Villach Alps; Schlotheim, Thuringia; J. F. W. von Charpentier, Saxony and the Erzgebirge; and Johann Karl Freiesleben (1774–1846), one of the most accomplished of the mineral geographers, reported on the Harz on Saxony, and on Mansfield-Thuringia. Elsewhere in Europe, von Buch was famous for his report on Norway and Lapland, Jens Esmark (1763–1839) surveyed Hungary and Transylvania in 1797, Jean de Charpentier produced a report on the Pyrenees in 1823, Berger worked on Ireland, Brochant de Villiers (1772–1840) carried out the official survey of France, Cuvier and Brongniart produced their influential memoir on the region round Paris, Berger and Webster (1773–1844) worked on southern England and Jameson surveyed Scotland (Berger 1811; Webster 1814; Conybeare and Phillips 1822; Davies 1969). Still further afield, Alexander von Humboldt studied the geognostic relations in Central America, von Buch visited the Canaries, the first geological map of the United States was published by the Wernerian William Maclure in 1809, and Robert Townson, who had produced a mineralogical map of Hungary in 1797, agitated for the Wernerian cause in New South Wales (Vallance and Torrens 1982). Gradually the original Wernerian concentration on the primary and transition rocks shifted, and, because of the way formations were distinguished, Wernerian gradualism was replaced by a theory of earth history interrupted by catastrophic revolutions.

The second branch of the Wernerian radiation, the causal school, headed by von Buch, Humboldt, and Hoff, was an extension of the older causal tradition (see chapter 8). These geologists remained committed to the study of contemporary processes, to the appropriateness of laboratory analogies, to the one-directional nature of many causal changes, and to the importance of mineralogy in understanding the earth's causal processes. They accorded an increasingly important role to heat in the earth's economy, recognizing that this, although consistent with certain themes in chemical cosmogony, was largely a result of Hutton's work.

The advocates of these two branches of Wernerianism were personal friends, familiar with one another's work; some, such as von Buch and Humboldt, worked in both branches. However, the two branches developed largely in conceptual isolation. In the early 1830s, Élie de Beaumont juxtaposed the two branches to produce a new account of earth history.

I call both these divergent subgroups Wernerian because both retain a very substantial portion of Werner's theory. The skeptic might hint that this use of the term *Wernerian* is so broad as to exclude nobody. That is not so. It excludes, for example, the tradition of English practical geology, which independently developed many of the same substantive techniques as the Wernerians without their theoretical underpinnings. It also excludes the Huttonians, who did not believe in a unidirectional history of the earth and who did not adopt the concept of a formation. But it does include, and is intended to include, a very large portion of the geologists in the early nineteenth century, just as the terms *Cartesian* and *Newtonian* include very large numbers of eighteenth-century scientists. I do not intend to cover up real differences between geologists I describe as Wernerian; in fact I shall spend much of chapters 7 and 8 exploring them. But the differences cannot be understood without the base of shared assumptions.

Werner's theory was disseminated through a network of institutional and social contacts. Werner taught many of the most important geologists of the early nineteenth century (Eyles 1964). Leopold von Buch, Alexander von Humboldt, Jean de Charpentier, the mineralogists Friedrich Mohs (1773–1839) and D. L. G. Karsten (1768–1810), and the crystallographer Christian Weiss (1780–1856) all studied at Freiberg. Freiesleben, d'Aubuisson de Voisins (1769–1841), Ernst von Schlotheim, Brochant de Villiers, Jens Esmark, Moritz von Englehardt (1779–1842), Jean-André Deluc, Robert Jameson, Franz Reuss (1761–1830), and Ami Boué (1794–1881) all visited the institution for longer or shorter periods.

At least five of those who studied with Werner—Reuss, Jameson, Brochant de Villiers, d'Aubuisson de Voisins, and A. M. del Rio—wrote textbooks (del Rio 1795–1805; Reuss 1801–6; Brochant de Villiers 1801–2; Jameson 1804–8; d'Aubuisson de Voisins 1819). Werner did not approve of what he thought of as a pirating of his notes. He complained, "A sort of mercantile speculation has been made of these manuscripts [lecture notes] in other countries" (Werner 1791/1809, xxviii). But since the authors treated every aspect of Werner's theory, including the description of the external characters of minerals, the theory of mineral classification, and geology ("geognosy"), they very effectively disseminated his ideas.

Even geologists who had not studied at Freiberg became familiar with Werner's theory.

Scholars who had not visited Freiberg produced further works that were straight explications of Werner, or that owed much to his influence. Richard Kirwan in Ireland, William Phillips, W. D. Conybeare and Robert Bakewell in England, and Parker Cleaveland in the United States, added to the texts already available in France, Germany, Scotland, and Mexico (Kirwan 1799; Bakewell 1813; Phillips 1815; Cleaveland 1816; Conybeare and Phillips 1822). I describe these texts as Wernerian because most covered the same topics in the same terms as the Wernerian texts already listed, even when their authors questioned some aspects of Werner's theory. Kirwan and Cleaveland wrote orthodox Wernerian texts. Bakewell followed the Wernerian sequence of the formations.

Phillips's textbook comprises two lectures on mineralogy, arranged in a Wernerian way, and three lectures on geology. The bulk of these three chapters is given over to a description of the sequence of formations and veins, described in Wernerian language. Concluding with a comparative description of the geological agencies invoked by Hutton and Werner, Phillips (1813, 185) says that he finds it "impossible not to give a preference to that theory which ascribes the formation of primitive rocks to the agency of water" on chemical grounds. By 1822, when Conybeare and Phillips published their textbook, they eliminated mineralogy (while insisting on its importance) and limited their detailed discussion of the sequence of the rocks to the *Flötz*, arguing that in the case of Britain the alluvial rocks had already been discussed by Webster, and the primitive and transition by Macculloch. But they attempted to fit the *Flötz* into the German sequence, to correlate them, and to tidy up the terminology. While arguing that many of Werner's opinions were being abandoned, Conybeare and Phillips (1822, xiv) conceded that he had "done more than any other individual to promote [geology's] career."

By the succeeding generation, the Wernerian origin of these textbooks was sometimes forgotten, since the authors did not always mention Werner by name. But a study of the content makes their Wernerian origin very obvious. As a consequence, most geologists in the first thirty years of the nineteenth century learned their geology from a Wernerian textbook.

Werner and his pupils also dominated the journals that specialized in mineralogy and geology. In 1788, the mining academy at Freiberg sponsored the publication of the first significant geological periodical, the *Bergmannisches Journal*, edited by two of Werner's students, A. W. Kohler and C. A. S. Hoffman. Although it ceased publication in 1815, the Freiberg school replaced it with an annual review in 1827. A few years later, in 1794, the École des Mines, which was dominated by Wernerians, began publishing the *Journal des Mines*, continued as the *Annales des Mines* from 1816. These were followed in 1801 by K. E. A. von Hoff's journal, *Magazin für die gesammte Mineralogie, Geognosie and mineralogische Erdbeschreibung*, and in 1807 by Karl von Leonhard's *Taschenbuch für die gesammte Mineralogie*. In Scotland, Jameson instituted the *Memoirs of the Wernerian Natural History Society* in 1811. The *Transactions of the Geological Society of London* began to appear in the same year and had a distinctly Wernerian flavor for the first few volumes. In 1818, two further journals were started, both strongly Wernerian, the *Transactions of the Royal Geological Society of Cornwall* and the mineralogist C. J. B. Karsten's *Archiv für Bergbau und Hüttenwesen*. With the exception of von Hoff's journal, which ceased publication after a short while because of the failure of its publisher, these journals all ran to several and often to many volumes. And, of course, there were more general scientific and scholarly journals that carried Wernerian work as well.

Wernerians controlled teaching institutions in geology every bit as much as they controlled the textbooks and journals. In Germany, Werner continued to teach until his death in 1817, and his influence remained strong long after. In France, Brochant de Villiers took an appointment at the École des Mines, where he had once been a pupil, and trained Élie de Beaumont, Pierre-Louis-Antoine Cordier, and O. P. A. Dufrénoy, among others. In Mexico, Fausto de Elhuyar and A. M. del Rio, both pupils of Werner, taught mineralogy at the mining school that was opened there in 1792 (Victor Eyles 1969). In Scotland, Robert Jameson taught Wernerian theory at the University of Edinburgh (Sweet and Waterstown 1967; Chitnis 1970). True, his reputation as a teacher has been damaged by Darwin's well-known remark that Jameson's lectures were so "incredibly dull" that he determined never "to read a book on geology or in any way study the science" (Barlow 1958, 9). But Jameson's other pupils, including Ami Boué, Edward Forbes (1815–1854), James Forbes (1809–1868), Richard Owen (1804–1892), Charles Daubeny (1795–1867), and William Scoresby (1789–1857) did not complain (Sweet in Jameson/1808 1976). Indeed, Jameson probably taught Wernerian theory to more geologists than did Werner himself. In England, John Kidd (1775–1851) and Daubeny taught Wernerian mineralogy at Oxford.

A good case can be made that even the English were but one part of the general Wernerian radiation. Following William Buckland, Nicholaas Rupke (1983) has recently distinguished an "English school" of geology, distinct from those of Freiberg, Paris, and Edinburgh. I would see it as one part of the Wernerian radiation, a school within the general tradition rather than a completely distinct movement. According to Rupke, the English school emphasized paleontological stratigraphy, the geology of their own region—namely, the (largely *Flötz*) geology of England and Wales—diluvialism, the interplay of geology and religion, particularly natural theology, and downplayed the economic utility of geological theory. But paleontological stratigraphy and regional geology, which constitute the core concerns of the English school, were typical of Wernerians everywhere. The interest in diluvialism and in the relations between geology and religion *were* particular to Britain. Werner, for example, although from a pietist background, appears to have given little weight to religious considerations in his geology. Indeed, in his student days, he even wrote a humorous poem, rebutting the charges of his friends with the response, "Wie könnt ihr mich noch Atheiste nennen?" (How can you still call me an atheist?) (Ospovat 1971, 26). Why the English were different is an interesting problem, but one that lies outside the province of this book (Gillispie 1959; Brooke 1979). And this, like their disinterest in practical matters, in no way diminishes the fact that their *theoretical* commitments were Wernerian.

For example, the Geological Society of London, just like its continental counterparts, began with a Wernerian orientation. It was founded by certain members of the defunct British Mineralogical Society, who had combined to finance the publication of the *Traité complet de la chaux carbonatée* by the French royalist emigré the Count de Bournon (Woodward 1908; Rudwick 1963; Laudan 1977b; Weindling 1979). The first president of the Society, George Bellas Greenough (1778–1855), and Humphry Davy succeeded in persuading the members that the society's aim should be to study "nature on the great scale"; therefore they chose the name "Geological Society" rather than "Mineralogical Society." The title was promissory, not descriptive, for only one of the thirteen founder members had published on geology as opposed to mineralogy. Insofar as they adopted any theory, they were almost without exception Wernerians. In a series of combative articles, the practical geologist John Farey, who came from an indigenous tradition, attacked what he saw as the overweening Wernerianism of the Geological Society. He wanted to promote "the cause of practical English geology against the theoretical pretensions of an Anglo-German Geognosy, which has too high and proudly raised its intolerant head amongst us" (Farey 1815, 333–34).

Thus, the English were as indebted to Werner as were the Germans and the French. But, with some exceptions, such as Robert Townson, who had studied with Sage at the École des Mines in Paris, and Thomas Weaver, who had been to Freiberg itself, they were hampered by their lack of training in the technical aspects of geology, such as mineralogy and crystallography. By continental standards, they were amateurs in the pejorative as well as the honorific sense.

Returning to the German states, we find that the Wernerian radiation extended beyond mineralogists and geognosists to German intellectuals of all kinds. Goethe subscribed to Wernerian theory until the end of his life, deploring the overthrow of Werner's account of the origin of basalt. Many well-known romantics, including Benedict von Baader, Heinrich Steffens (1773–1845), Gotthilf Schubert (1780–1860), Theodor Koerner, and Friedrich von Hardenberg (Novalis), studied with Werner and retained their interest in mineralogy and geognosy (Ospovat 1982). Novalis worked briefly for Saxon geological survey; Steffens taught geology and mineralogy at Halle, Breslau, and Berlin; and Schubert taught at the universities of Erlangen and Munich.[2] The coalescence between Wernerianism and romanticism is well illustrated by von Buch's address (1808) on his election to the Berlin academy of sciences. In this he abandoned his usual sober style and postulated a comprehensive theory of the development of the earth and the objects on it. He argued for the interplay of natural forces that unified everything from the crystallization of granite to the strivings of the human intellect.

The source of this appeal is not entirely clear. True, both closeness to nature and the hint of religious and political freedom [Freiberg] had been associated with mining and geology since the opening of the mines of the Erz Mountains in the fifteenth century. Whatever the reality, German intellectuals found this intriguing. Perhaps more important, they saw the Wernerian account of the history of the earth as extending backward in time the histories written by political and cultural historians. Interest in history in late-eighteenth-century Germany was intense. By 1790 there were more than one hundred predominantly historical journals in the German-speaking states, and it has been estimated that, of the 5,000 writings published in the three years prior to 1772, one-fifth were historical (Butterfield 1955). The new historical school centered on Göttingen gave pride of place to universal history and insisted that history be based on a critical interpretation of original sources. Werner's geology pushed back before the appearance of man on earth, serving as the backdrop to the cultural history that could be written for the period of man's habitation. His insistence on an examination of the rocks was the geological equivalent of the historian's examination of the texts.

A much more detailed analysis of German thought in the period would be required to understand fully the role played by Werner's geognosy. But for our purposes it is enough to note that some day his appeal to such a wide range of interests needs investigation.

Those familiar with the history of geology will recognize that this account differs from the standard picture of Werner. Despite the efforts of a number of scholars, particularly Alexander Ospovat and East German historians, such as Hans Baumgärtel, Martin Guntau and Otfried Wagenbreth, Werner still suffers from a bad press. The author of one of the most popular books on geology published in recent years, John McPhee (1980, 93–94), dismisses Werner as "a German academic mineralogist who published very little" but who carried "forward to the nth degree" the theory that rocks had been formed in water, being "gifted with such rhetorical grace that he could successfully omit such details" as where the water went. Werner, he claims, was the "lineal antecedent of what has come to be known as blackbox geology— people in white coats spending summer days in basements watching million-dollar consoles that flash like northern lights."

McPhee is simply summarizing a venerable Anglo-Saxon historiography that can be traced back to the history of geology with which Lyell prefaced his *Principles of Geology* (1830–33) (Ospovat 1976). Lyell took his account of Werner from Cuvier's *éloge*, which, in turn, was based on Humboldt's reminiscences about his mentor (Coleman 1963). Lyell passed over most post-Wernerian geology, distancing his own work from that of his immediate predecessors. Perpetuated in numerous popular and scholarly accounts of the history of geology, this historiography presents Werner in a most unflattering light (Geikie 1905/1962). He is accused of having failed to carry out any fieldwork and of being an "armchair geologist," of having held a hopelessly simplistic theory of a universal ocean that laid down identical rocks in onionskin layers all round the earth (with the insinuation that this owed more to a fundamentalist interpretation of the Bible than to scientific investigation), and of bringing so many young geologists under the sway of his persuasive lecturing style that he retarded the progress of geology for a generation. It is high time these myths were laid to rest.

6

The Huttonian Alternative

In 1788, the Scottish natural philosopher James Hutton (1726–1797) published "an investigation of the laws observable in the composition, dissolution, and restoration of land upon the globe" (Hutton 1788; see Eyles 1950 and 1955 for publication history). Hutton's aims, methods, and substantive theory of the earth were quite foreign to mineralogists in the Becher-Stahl tradition. His theology led him to construct a cyclic causal theory of the earth. By contrast, those geologists who were not indifferent to the possible relevance of theology to geological theorizing accepted a Christian theology that bolstered a genetic or historical theory of the earth, not a cyclic one. In this they were of one mind with the dominant tradition of chemical cosmogony. Hutton relied heavily on the methods of hypothesis and eliminative induction at a time when his contemporaries thought them less reliable than the methods of analogy and enumerative induction. He constructed his theory of the earth in terms of a Scottish interpretation of Boerhaavian heat theory at a time when his contemporaries were almost all indebted to Stahl. In short, Hutton worked in an intellectual context quite different from most mineralogists.

Hutton is best understood in the setting of the Scottish Enlightenment. He had received his medical degree from the University of Leiden in Holland in 1749. Returning to his native Scotland, Hutton led the life of an improving landlord for a number of years, experimenting with new methods of cultivation and investing in a chemical works, activities that were commonplace among the more forward-looking gentry of both England and Scotland in the latter half of the eighteenth century. Perhaps tiring of the rural life, Hutton took up residence in Edinburgh in 1768, where he stayed for the remainder of his life. Here he enjoyed the companionship of the chemist Joseph Black and Adam Smith (1723–1790), best known for his treatise *On the Wealth of Nations*. They and a number of other scholars, such as Adam Ferguson, John Playfair, and James Hall, met regularly at the Royal Society of Edinburgh, a society dedicated to the discussion of topics in natural philosophy. It was to this group that Hutton first publicly presented his theory of the earth (Donovan 1977).

Some years later, in response to criticism from Richard Kirwan and others in the Wernerian school, Hutton extended his original paper, published in 1788, to two long volumes—the *Theory of the Earth, with Proofs and Illustrations* (1795). In the first chapters, he restated the original theory; in the remainder of the work, he attempted to provide empirical support from his own observations and an extensive reading of the geological literature. Hutton's theory was widely noted, not just in Britain but on the Continent (Anonymous 1788a, 1788b; Dean 1973). Desmarest, for example, published Hutton's 1785 sketch and parts of the 1788 theory in the first volume of the *Géographie physique* (1794/5–1828).

Like Werner, then, Hutton proposed a comprehensive account of the earth in the 1780s, an account that became widely disseminated at the beginning of the nineteenth century. But there the similarity ends. Unlike Werner, Hutton did not attract a major following, except possibly in his native Scotland. To understand why this was so, I shall present his theory as his European and English contemporaries, already familiar with the chemical cosmogonical tradition stretching from Becher to Werner, would have read it. Seen in this light, the idiosyncrasies of Hutton's theory are evident. That done, I shall go on to explicate the theological, natural philosophical, and chemical underpinnings of Hutton's theory in order to explain how he had come to formulate a theory so unlike Werner's. I shall conclude by examining the reaction to Hutton's theory.

THE HUTTONIAN THEORY OF THE EARTH

"This globe of the earth," declared Hutton (1788, 209–10), "is evidently made for man. . . . [It] is a habitable world; and on its fitness for this purpose, our sense of mission in its formation must depend." No mere rhetoric, his explicitly teleological vision of the purpose of geological theorizing was fundamental to his work (Galbraith 1974; Grant 1979). "The proper purpose of philosophy," Hutton (1792, 262) maintained, "is to see the general order that is established among the different species of events, by which the whole of nature, and the wisdom of the system is to be perceived."

Such congruence between teleological concerns and theories of the earth was not unusual in eighteenth-century Britain, though much rarer on the Continent. But Hutton differed from other geologists who wished to reconcile their science and their religion. Like a number of his friends in the Scottish intellectual community, he was not a Christian but a deist. The difference is crucial.

As a deist, Hutton asserted the existence of a deity, but denied the evidences of revealed religion. In particular, like other deists, he dismissed the Christian story, laid out in the Old and New Testaments, as unworthy of serious consideration. Evidence of the existence and nature of God was to be found in the natural world, not in revelation. Christians sought to reconcile the evidence of revealed religion with the evidence of the natural world. In the eighteenth century, many Christian geologists thought that the Bible, like other ancient writings, offered invaluable direct testimony about the earth's past. Deists had no use for such endeavors. Biblical stories were mere myths, not to be taken seriously by the natural philosopher. Hutton's deism underpinned his theory of the earth. Readers could be in no doubt about Hutton's theology.

Theological differences between Christians and deists determined not only the evidence they considered relevant to theories of the earth but the very *kind* of theory that they advanced. Christians were committed to unidirectional theories. They stressed the earth's original formation, were receptive to evidence of dramatic and singular events that altered the face of the earth and that could be assimilated to the biblical Flood, and assumed that the earth would come to an end at some point in the future (see, e.g., Burnet 1681/1961; Deluc 1809; Buckland 1823).

Deists, unlike Christians, were free to consider other patterns of earth history. They did not have to restrict themselves to unidirectional theories, though some such as Werner, who appears to have been closer to deism than to any other theological position, certainly did. Hutton, by contrast, did not believe that the deity would have created a world that required divine intervention for the maintenance of habitability. Rather, the deity's perfection would be mirrored in a natural world that was so designed and balanced that it could maintain indefinitely a surface suited for God's creatures. Hutton's avowed aim in his theory of the earth was to explain how habitability was maintained by explicating the laws of nature on which it rested. Unlike Christians, who thought that God had designed a world with a distinct beginning and a definite end, he hypothesized a world that, as far as the geologist could tell, extended indefinitely far into the past and would last indefinitely long into the future. Underlying the globe's apparent disorder, according to Hutton, was a system nicely adjusted to ensure the continued maintenance of its habitable surface. On this earth "we find no vestige of a beginning,–no prospect of an end" (Hutton 1788, 304; see also Playfair 1802). Such an unabashed incorporation of theological concerns into the very form of a theory of the earth was foreign to most of his contemporaries in Europe.

Given Hutton's aim of accounting for continued habitability, what system did he propose? He began by stating that a necessary precondition was that the earth's position in the solar system remain stable, ensuring constant physical conditions on the earth. To Hutton's mind, Newton had already solved this problem. A combination of gravitational forces and the inertia (*vis insita*) of moving bodies guaranteed stability. Taking Newtonian theory for granted, Hutton used it as a model to solve the more specifically geological problem—namely, how the seemingly chaotic and disorderly surface of the earth was maintained in a habitable state.

Hutton postulated a cycle of decay and renovation (parallel to the cyclic motions of the heavens) that would sustain the globe overall (though not any particular spot on it) in a state fit for man and other creatures (Dott 1969).[1] Rocks were eroded by water, and their remains washed down to the sea. There renovation of land began. New rocks were consolidated from the remains of the old. Then they were elevated to form new land, completing the cycle. But there was an important difference between the earthly cycles Hutton postulated and those observed in the heavens. The latter were readily observable; the former were not. Hutton's renovation phase occurred in regions of the globe inaccesible to human observation and on a time scale that dwarfed the human life span. Geologists could only observe the rocks now at the earth's surface. They had to construct a chain of reasoning from the observable to the unobservable, from the earth's surface to its interior, and from the present to "the natural operations of time past" (Hutton 1788, 218). Hutton's reasons for *wanting* a cyclic theory of the earth were quite clear. But his readers, who by and large did not share the theology that supplied these reasons, would have waited to see what *evidence* Hutton offered for the occurrence of cycles.

Hutton began to offer such evidence by arguing that the earth's surface was continually destroyed to provide renovating materials.

Surface rocks, he said, must be broken down into soil in order to support the plants that sustain other forms of life. Many geologists, including the Wernerians, agreed that such destruction occurred, though the agreement was not necessarily on teleological grounds (see chapter 5; also see Davies 1969). They drew a variety of conclusions about the implications of this destruction. Some, such as Jean André Deluc (1727–1817), the prolix emigré Swiss geologist and friend of Blumenbach, argued that the soil protected rocks from complete breakdown (Deluc 1790). Some, such as John Williams (1789), the regional English geologist, thought that denudation was beneficial if not too extensive. Few contemplated the complete destruction that Hutton proposed. (Hutton's interesting theory of erosion, which Playfair extended, has limited bearing on the main themes of this book, and I shall not pursue it here.)

Hutton's theological bent meant that he pondered the destruction of the land with equanimity. Repair would necessarily follow, he thought. If it did not, then either "the system of this earth has . . . been intentionally made imperfect" or "it has not been the work of infinite power and wisdom"—two options that Hutton regarded as self-evidently impossible (Hutton 1788, 216). We should not think of the earth as a machine, he suggested, but as an "organized body [that] has a constitution in which the necessary decay of the machine is naturally repaired." Once more, his readers would have wanted evidence for this opinion.

Hutton's key evidence lay in the limestones and marbles (the "calcareous earths"). Reportedly a keen collector of minerals, he accepted the standard fourfold classification of minerals into earths, metals, salts, and sulfurs. Hutton (1788, 218) asserted that the calcareous rocks were *entirely* composed of the shells of marine creatures—"calcareous bodies which had belonged to animals." Since this was not the common view, Hutton suggested some reasons why his readers should accept it. Many marbles and limestones contained shell and coral visible to the naked eye. Others contained shell and coral mixed with crystals. And yet others were fully crystalline. A continuum existed from shelly to crystalline calcareous rocks. Hutton put the gradation down to the extent to which the shells had been rendered fluid and lost their original form when the rock consolidated. He concluded that all calcareous rocks were the recycled remains of an *inhabited* earth. This, of course, was crucial to his overall aim of showing the sustained habitability of the earth. By analogy, Hutton (1788, 245) concluded, "All the strata of the earth, not only those consisting of such calcareous masses, but others superincumbent upon these, have had their origin at the bottom of the sea." Consequently, some shift in the relative positions of the new rocks and the surface of the ocean must have created new land surface.

Having presented his argument for a cycle of destruction, consolidation, and elevation, Hutton considered its causes. The repair of the globe, said Hutton (1788, 223), required

> an operation by which the earth at the bottom of the sea should be converted into an elevated land [and] a consolidating power, by which the loose materials that had subsided from water, should be formed into masses of the most perfect solidity, having neither water nor vacuity between their various constituent parts, nor in the pores of those constituent parts themselves.

Like the Wernerians, Hutton (1788, 224) insisted that consolidation could be understood only with "the science of chemistry," which studied "the changes produced upon the sensible qualities, as they are called, of bodies." Consolidation was a two-phase process: the earth at the bottom of the sea had first to become fluid and lose its original form; then it was consolidated in a new form. Thus far this is familiar. But Hutton was using a version of the science of chemistry very different from Stahl's.

Hutton thought there were two ways a body could be consolidated—by accretion (his term for what most chemists called concretion) or by congelation (a term that he shared with other chemists). The former took place "mediately with the assistance of a solvent," which had to be water in the case of geological processes (Hutton 1788, 225). The latter was brought about "immediately by the action of heat."

Hutton dismissed accretion, and with it the efficacy of water as an agent of consolidation. According to Hutton (1788), and contrary to the perception of the Wernerians, chemists had demonstrated water's inability to dissolve many minerals, including the rock-forming siliceous and sulfureous minerals. Furthermore, water could not easily pass through the body being solidified. For this to happen, "water must be, like fire, an universal solvent, or cause of fluidity, and we must change entirely our opinion of water in relation to its chemical character" (Hutton 1788, 229). Thus, for the water to escape after consolidation, a "separatory operation" would be required. Not seeing how this could occur in the bowels of the earth, Hutton concluded that, if water were the agent of consolidation, we should expect to find water in rock pores. Since we do not, water could not be the cause of fluidity and consolidation.

But if heat or fire were chosen as the agent of fluidity, all the difficulties vanished:

> The loose and discontinuous body of a stratum may be closed by means of softness and compression; the porous structure of the materials may be consolidated, in a similar manner, by the fusion of their substance; and foreign matter may be introduced into the open structure of strata, in form of steam and exhalation, as well as in the fluid state of fusion. [Hutton 1788, 229–30]

Rocks made from all the different mineral classes—the calcareous, siliceous, and other earths, the metals, the sulfurs, and the salts— must have been consolidated by heat (Hutton 1788). Heat penetrated the globe and fused the loose materials on the ocean floor in the process. Heat was subtle enough to escape completely as congelation took place, and the resulting rock was completely solid without any residual foreign matter. Hutton left the details vague, convinced that his readers would accept consolidation by heat as the only possibility once they appreciated the force of his arguments against water as the consolidating agent.

In sum, Hutton believed *all* rocks had been rendered solid by heat, not water. He must have realized that his contemporaries would find this hard to accept, especially for those rocks like the marbles and limestones, which were known to disintegrate to a powdery calx on heating. He did point out that many calcareous strata contained "flints." Since flints are glassy nodules, any eighteenth-century mineralogist would have agreed that their glassy texture was evidence that they had cooled from a melt. They would have been less ready to accept that stony or crystalline minerals had been consolidated by heat. Later, in response to criticism, Hutton (1795, 65–66) invoked the principle of compression. Under the immense weight of the ocean, he argued, volatile gases would not be able to escape, and the consolidation that he described would indeed take place. "No further conditions are required, than the supposition of a sufficient intensity of subterraneous fire or heat, and a sufficient degree of compression upon those bodies, which are to be subjected to that violent heat, without calcination or change." The effect of compression on chemical changes brought about by heat was "not a matter of supposition [but] . . . an established principle in natural philosophy" (Hutton 1795, 94). He claimed that experiments (that he neither cited nor described, although he did mention that they were difficult to perform) showed that increasing pressure decreased the separation of the volatile parts from a heated substance.

Hutton thought that granite, like other rocks, was consolidated by heat, but at first he was not sure whether it was originally stratified or had been injected from the subterraneous regions. He toyed with the idea that, as most geologists believed, it was a primitive rock. He suggested a thought experiment in which the earth's strata were redistributed over the sea bottom. The result would be "a spheroid of water, with granite rocks and islands scattered here and there" (Hutton 1785, 222). But for his system of continued habitability, he wanted to show that granite was not prior to the stratified, recycled rocks, but that it was part of the same cycle of decay and renovation. Therefore, he set off for Glen Tilt in 1785 to see if he could find evidence that granite had been intruded into strata. Finding veins of granite cutting across the overlying strata, Hutton (1790, 81) decided that it was not "the original or primitive part of the earth." So excited was he with this empirical evidence in support of his theory of the recycling of the rocks, that "the guides who accompanied him were convinced that it must be nothing less than the discovery of a vein of silver or gold" (Playfair 1805, 68–69).

Finally, there was the question of the appearance of new land above the ocean. Hutton proceeded just as he had done with the question of consolidation. Adopting the method of eliminative induction, he outlined what he took to be two mutually exclusive and exhaustive hypotheses—namely, that new land appeared either as a result of the retreat of the ocean or as a result of the elevation of the sea floor. He then rejected the first, assuming that the reader would accept the remaining alternative. Once more, the alternative that he rejected was the one commonly accepted by geologists. There were two reasons, he thought, why new land could not have appeared as a result of the retreat of the ocean, as the Wernerians had suggested. First, no one could say where the water had gone. Second, no one could say where the continent that had supplied the materials for the new land had gone.

Therefore, Hutton concluded that the ocean bottom must have been elevated. Once more, Hutton turned to heat. The "power of heat," Hutton (1788, 263) claimed, had an "unlimited" ability to expand bodies. He anticipated that potential skeptics might ask how heat or fire could operate at the bottom of the ocean, but did not give a causal explanation. He simply asserted that, in his discussion of consolidation, he had already shown that heat did act in this way.

Hutton offered some "proofs" of his hypothesis. First, the strata exposed at the earth's surface were inclined at all angles to the horizontal. Unlike the mineralogists in the Becher-Stahl tradition, who assumed this was the original position of the strata, Hutton assumed they had been deposited horizontally and then disturbed. Second, the strata were broken up, although they must have been deposited as continuous beds, and they were folded, although they must have been deposited on a plane surface. Veins, too, must have been formed during uplift. "Ask the miner," asked Hutton (1788, 266–67) rhetorically, "from whence has come the mineral into his vein?" Ignoring the answer that German miners would have given to this question, Hutton argued that the form and content of veins (excluding those that had been formed during consolidation) showed that they were "some of the expanded matter . . . condensed in the bodies which have been heated by that igneous vapour."

Finally, he insisted that volcanoes showed that heat was operating with undiminished vigor. They acted as safety valves to prevent the continued action of the expanding force in areas where continents had already been elevated. Erosion would then begin once more, and a new cycle would be inaugurated. "These operations of the globe," Hutton (1788, 272) said, "remain at present with undiminished activity, or in the fullness of their power."

THE IMMEDIATE REACTION

Hutton's theory, as I have said, did not go unnoticed. But it was not accorded a warm welcome. Numerous reasons have been advanced to explain this, including his bad prose style (Playfair 1802), his unorthodox theory of erosion (Davies 1969), his equally unorthodox theory of heat (Gerstner 1971), and most generally his theology (Gillispie 1951/1959; Dean 1973). All these factors undoubtedly contributed. As should be clear, Hutton's theory flew in the face of all the conventional wisdom about geology that had been slowly built up in the eighteenth century. Kirwan (1793, 71) must have spoken for most of his contemporaries when he said that, while problems besetting the Wernerian system might "escape our reason," the problems besetting the Huttonian system were enough to "contradict" the best mineralogical and geological knowledge of the day.

In addition to this wide range of problems with Hutton's theory, all his critics focused on his theory of consolidation by heat, a theory quite at odds with the majority opinion about chemical processes. Kirwan (1793, 71) was particularly scathing about "the supposed igneous origin of stony substances. . . . The supposition of a degree of heat under any given compression, sufficient for the fusion of stony substances in general, without calcination or change, is not only gratuitous, but contrary to all that we at present know of the agency of heat." The calcareous rocks, which, as we have seen, were so central to Hutton's system, presented the worst problem as they turned to a powder, not a rock, on heating. Even Hutton's friend (and mentor where chemistry was concerned), Joseph Black (1755, 40), *defined* the calcareous earths as those "that are converted into a perfect quick-lime in a strong fire." Heating limestone, to produce the powdery calx, quicklime, was one of the simplest and best known procedures in industrial chemistry. All over the limestone outcrops of the British Isles and Europe, small limekilns were to be found where limestone was heated at moderate temperatures in order to produce quicklime for mortar and fertilizer.

Worse still for Hutton, those who believed in consolidation by water could easily explain the consolidation of limestone. Compared with many other minerals, calcareous earth was relatively easily soluble in water, particularly in slightly acidic water. In the caves and springs of limestone regions, tourists admired the growth of stalactites and stalagmites and dipped crutches, walking sticks, and a motley collection of other objects in the water to see them become coated with a hard covering of limestone within a month or so. This limestone was granular, but in more tranquil conditions, the "sparry" (crystalline) form could also be produced.

Hutton may well have accepted Black's theory that the loss of "fixed air" (carbon dioxide) from a heated calcareous earth was not a significant chemical effect though he did not mention it. Black (1755, 22) had argued that heated calcareous earths suffered "no other change in their composition than the loss of a small quantity of their water and of their fixed air." But Black minced no words about the change in appearance; the calx was powdery, not stony.

John Murray, in his anonymous and admirably well-balanced *Comparative View of the Huttonian and Neptunian Systems of Geology* (1802, 28) claimed that the calcareous rocks provided a "direct demonstration of the falsity of the Huttonian hypothesis [that] the materials which are collected at the bottom of the ocean are at great depths exposed to the action of an intense heat, under a strong pressure, by which they are fused and consolidated." Deluc (1809, 350) agreed that they constituted "the most direct argument against that part of Dr. Hutton's theory, which supposes an excessive *heat* in the interior of the globe." Kirwan (1793, 71) pointed out that experiments carried out by Lavoisier and others had shown that the calcareous earths were "absolutely infusible in any degree yet known." Even the chemist James Hall (1761–1832), later to become one of Hutton's most loyal disciples, found so much of Hutton's theory "apparently in contradiction to common experience" that he reported that a first reading "induced [him] to reject his system entirely" (Hall 1812, 74).

The problem extended beyond the calcareous rocks, of course. Even if Hutton had been able to prove that there was some kind of heat or fire within the earth, and even if he could have shown that it was powerful enough to fuse all the rock-forming minerals, cooled mineral melts were generally thought to be glassy or powdery, not, like most rocks, stony or crystalline. Most mineralogists were not familiar with the new evidence that some melts could cool to form crystals and stones (see chapter 3). Thus, Hutton was worse off than Werner. Wernerians could show that water, under certain conditions, could dissolve or suspend all the rock-forming minerals, including the silicates, and deposit them as stony or crystalline solids. Hutton could not show that heat could accomplish this.

Many recent philosophers of science have claimed that theories are never rejected because of empirical anomalies (Kuhn 1962; L. Laudan 1977; Lakatos 1978). Hutton's critics would not have agreed. The anomaly of the consolidation of rocks, particularly the calcareous earths, was enough for them to reject Hutton's theory, cutting as it did at the core theses of his system. Except for some vague remarks about the principle of compression, Hutton had made no attempt to confront the problem in his geological works. Yet his use of heat as a cause for consolidation and elevation was one of his chief innovations. Since his theory of consolidation ran counter to the chemistry of the day, Hutton and his followers had either to reform chemistry or to show that the supposed empirical anomaly was chimerical. Either way, Hutton's theory of consolidation had to be brought into line with the increasing body of knowledge about the substances making up the mineral kingdom. Those geologists who, for one reason or another, thought that the Huttonian theory held more promise than the reigning alternative—the theory of consolidation by water—had no option but to try to offer some way out of these apparently crushing difficulties.

THE BACKGROUND TO HUTTON'S THEORY

Yet Hutton's system was far from arbitrary, as it must have seemed to the reader who looked only at the successive versions of his theory of the earth. Like Werner, who also relied on a theoretical infrastructure that he did not make explicit in his geological works, Hutton had a whole set of underpinning theories. But whereas Werner could count on most eighteenth-century mineralogists to recognize his presuppositions, Hutton could not. The purpose of the following exegesis is to reconstruct the missing background using Hutton's extensive writings on metaphysics, epistemology, and the physics and chemistry of heat (Hutton 1792, 1794a, 1794b; O'Rourke 1978). I shall argue that Hutton's theory was shaped by three influences quite foreign to the traditions of Becherian mineralogical cosmogony and botanical taxonomy that shaped Werner's theory. The first of these was his deistic theology, which, as we have seen, he had made explicit and which I shall not explore further here. The second was a native British tradition stemming from the work of Newton that explored ideas about matter and power. And the third was the theory of heat and chemistry articulated by Boerhaave (1668–1738) and modified by Black (Donovan 1978).

Hutton's theory of the earth depended on the balanced interaction of attractive and repulsive "matter" or "power." His concept of "power" was indebted to a long tradition of pondering the implications of Newton's remarks about the possibility of an aether in the list of queries appended to the second and later editions of the *Opticks* (Cohen 1956; Schofield 1970; Heimann and McGuire 1971; Christie 1981). Following Newton's lead, eighteenth-century natural philosophers had invoked "aethers" or "imponderable fluids" to explain puzzling phenomena in fields as diverse as optics, electricity, magnetism, physiology, chemistry, and heat theory. These fluids were assumed to be composed of mutually repelling particles, this property being referred to as "elasticity." They were also regarded as "subtle" in that they were supposed to be able to penetrate between the particles of ordinary bodies; and they were "imponderable" in that they had no detectable weight. Causal interactions between matter were attributed to these fluids. Over the course of the eighteenth century, the distinction between matter and the fluids or aethers became increasingly blurred for many scientists, until some, such as Hutton, dissolved the distinction almost entirely. We cannot be sure exactly who influenced Hutton, since he does not acknowledge his sources, but his dependence on eighteenth-century natural philosophy is clear.

"Powers," according to Hutton, were simply one way of conceiving of matter. Matter was defined in terms of its intrinsic *activity*, or as he put it, "matter may . . . be considered as acting powers"

(Hutton 1792, 501). "Matter" was not to be confused with everyday solid bodies. The bodies that we observe are, Hutton would have it, nothing more than the perceptual effects of these underlying powers. As Playfair (1805, 75–76) summed up Hutton's position, "The supposed impenetrability, and of course the extension of body, is nothing else than the effort of a resisting or repulsive power; its cohesion, weight, etc. the efforts of attractive power; and so with respect to all its other properties."

Hutton believed that there were two kinds of power or matter—gravitational matter, and matter that derived from the sun, or the "solar substance." Both were active, not passive. Gravitational matter acted by attraction; solar matter, by repulsion. The interplay of these two active matters (powers) accounted for all natural phenomena.

Gravitational matter could not be separated from bodies and was responsible for their hardness and solidity, two properties of great concern to geologists. It manifested itself in various ways. The attractive force was responsible for the tendency of bodies to move toward each other; the cohesive force was responsible for the movement of parts of an individual body toward their common center; and the concretive force was responsible for short-range interactions, including consolidation (Hutton 1792). The interaction of the attractive force with the *vis insita* kept the solar system as a whole, and the earth within it, in a stable configuration. The cohesive and concretive forces consolidated fluid material on the sea bottom. But unless balanced by repulsive (solar) matter, gravitational matter would have collapsed to create an inert universe.

According to Hutton, solar matter, particularly in the form of heat, was the second great power of the universe, responsible for activity in the globe. The solar substance (as "latent" heat) rendered the loose material on the sea bottom fluid prior to its consolidation. And the solar substance (as "sensible" heat) combined with the newly consolidated bodies, causing them to expand and elevate. To understand this, we have to look at Hutton's predecessors.

There is indirect, but persuasive, evidence that Hutton's theory of heat owed a great deal to Boerhaave, as well as to Black. Hutton had studied at Leiden when the medical curriculum was still largely as Boerhaave had left it, and Hutton's good friend Joseph Black was also much impressed by Boerhaavian theory. Boerhaave was a figure to be reckoned with in the mid-eighteenth century (Metzger 1930). Like those in the British Newtonian tradition, Boerhaave was indebted to Newton, so Hutton's sympathy with his theory is part of a general pattern. Boerhaave's fullest discussion of heat occurred in his *Treatise on Fire*. The reputation of this work is summed up by the comment by Macquer (1766/1771, xi), in his widely distributed and translated *Dictionary of Chemistry* that it was "an astonishing masterpiece, so complete, that the human understanding can scarcely make an addition to it."

For Boerhaave, the ultimate source of heat or solar substance was the sun. It transmitted the solar substance to the earth as light, which penetrated bodies there as heat. Boerhaave vacillated between various hypotheses about the constitution of fire, at times suggesting that it was simply the motion of elementary particles, at others theorizing that it was a subtle substance that could penetrate the pores of bodies. He was sure that it could not be isolated in pure form.

Boerhaave used the solar substance to explain a wide range of phenomena, including physiological processes and combustion, in both cases integrating it with a form of the phlogiston theory. But for Hutton, the part of Boerhaave's theory that was useful in constructing a theory of the earth was his theory of the efficacy of heat in causing both the fluidity and the expansion of bodies. Increasing the amount of heat fluid in a body caused expansion or "dilation." The extent of this expansion varied from body to body, each body having its own characteristic rate. An increase in heat—that is, in Hutton's terminology, an increase in the amount of the "solar substance"— could frequently be detected by the increasing warmth of the body to the touch, though a more accurate method was to use a thermometer.

Furthermore, fluidity in bodies resulted from their contained heat. The addition of heat to a body rendered it fluid; removal of heat caused congelation. This applied to water as much as to any other substance. Water was fluid simply because it contained heat. Remove the heat, and the water congealed to ice. On this account the sharp distinction between consolidation by heat and consolidation by water vanished. Contained heat was the source of water's ability to dissolve other substances. Lacking that, water ceased to be a solvent.

Black elaborated on Boerhaave's account of heat (Donovan 1975). He distinguished two forms of heat, one primarily responsible for expansion, the other for fluidity. "Sensible" heat—that is, heat that we can sense—caused expansion. "Latent" heat—that is, heat that can not be immediately detected by the senses—was responsible for fluidity. Black's biographer quotes him as saying,

Heat is in nature the principle of fluidity and evaporation, though, in producing these effects, it is latent in respect to the thermometer, or any sensation of ours; and as matter, otherwise quiescent, becomes voluble and volatile in liquid and in vapour, heat may be considered in nature as the great principle of chemical movement and life. If it passes through vacuity as well as through body, as it certainly does in its communication from the sun to the planets, we must consider it not as an accident in bodies, but as a separate and specific existence, not less so than light or electric matter. [Ferguson 1805, 107]

Latent heat, according to Black, became chemically "fixed" in bodies in the same way that "fixed air" (carbon dioxide)—another elastic fluid—became fixed. Fixation caused these fluids, whether heat or air, to lose their usual properties. Fixed heat was no longer perceptible to the touch, but rendered bodies fluid by combining with gravitational matter. Further addition of heat caused the fluid body to expand; removal of heat caused it to congeal to a solid (Donovan 1975).

Hutton dedicated his *Dissertations on Different Subjects in Natural Philosophy* (1792) to Black. Comparing Black's work on heat to Newton's work on gravity, he eulogized Black's discovery of "*Latent heat*,—the Principle of Fluidity,—a Law of Nature most important in the constitution of this World,—and a Physical Cause, which, like Gravitation, although clearly evinced by science, is far above the common apprehension of mankind."

Hutton, with Black's knowledge, applied his modifications of Boerhaavian heat theory to the theory of the earth. In his biography of Hutton, Playfair (1805, 95–96) reported, "In every material point of philosophy [Hutton and Black] perfectly agreed. The theory of the earth had been a subject of discussion with them for many years, and Dr Black subscribed entirely to the system of his friend." True, Black may have had qualms about some of Hutton's theses, for Playfair wryly added, "Dr Black dreaded nothing so much as error, and . . . Dr Hutton dreaded nothing so much as ignorance."

The solar substance, emanating from the sun as light, combined with bodies on earth (Hutton 1792). It overcame their tendency to collapse into inert masses under the action of gravitational power. As latent heat, it rendered bodies fluid prior to consolidation on the ocean floor; as sensible heat, it caused rocks to expand, elevating land above the sea. From Hutton's perspective, heat was one of the two counterbalanced powers that maintained the economy of nature in a stable system. For geologists to attribute the major changes on the globe to the action of water was folly to Hutton, the result of confusing an epiphenomenal cause with a basic one.

This background helps us understand why Hutton constructed his sytem as he did. Even so, its details remain frustratingly vague. Why should the incoming solar substance affect deposits at the bottom of the ocean, but not under the continents? Or did Hutton think heat rendered granite, say, fluid under the continents? Why should the attractive, gravitational powers overcome the repulsive, solar powers to change fluid bodies into solid rocks? Did the expansive power of heat cause elevation directly, or did it work indirectly through the injection of molten matter? Hutton does not supply answers to these questions. Small wonder, then, that the majority of geologists who did not understand, let alone share, his general conceptual framework found the system so unsatisfactory. But before turning to the debates about his system, we should briefly consider Hutton's methodology.

A long historiographic tradition, not yet entirely defunct, has represented Hutton as an inductivist, who, unlike his rival Werner, reported his observations as he saw them without drawing unwarranted theoretical inferences and thus put geology on a scientific footing. Before Hutton, it has been asserted, "geology did not exist" (MacIntyre 1963, 2; see also Gillispie 1951/1959; Bailey 1967). "Instead of invoking conjecture and hypothesis, [Hutton] proceeded from the very outset to collect the actual facts" (Geikie 1905/1962, 299). His work was marked by "its resolute refusal to speculate on origins, and a rigorously empirical approach" (Gillispie 1951/1959, 46; see also Tomkieff 1947).

Yet in the 1960s and 1970s a wave of scholarship, to which my account is indebted, discredited once and for all the notion that Hutton eschewed speculation and placed him firmly within the cosmogonic tradition (Gerstner 1968; Davies 1969; Dott 1969; Heimann and McGuire 1971; Porter 1977; Grant 1979). As we have seen, his theory of the earth was shaped by his theories about the nature of god, knowledge, and matter.

Does this mean that Hutton failed to appreciate the importance of evidence? Not in the least. Like most of his contemporaries in Scotland, Hutton was deeply concerned with the nature of scientific method (Olson 1975). He favored the method of eliminative induction and, when he could not achieve that, the method of hypothesis. In his commitment to the latter, he was out of step with the largely inductivist temper of late-eighteenth-century science, although he was not alone (see chapter 1). Both the method of eliminative

Fig. 8. Hutton's illustration of the effects of elevation. John Clerk's 1787 sketch to illustrate Hutton's observation at Jedburgh, Borders (Craig ed. 1978, 57). This drawing was the basis of the engraving for plate III in Hutton's *Theory of the Earth* (1795, vol. 1). Horizontal beds of sandstone and puddingstone overlie vertical greywackes and shales, which Hutton believed were responsible for elevation.

induction and the method of hypothesis required the formulation of speculative theories. And both of them required data for the assessment of those theories. Hutton took the search for this data very seriously indeed. He traveled widely in search of new geological data. He had his observations recorded with detailed and accurate sketches (Oldroyd 1971; Craig 1978) (fig. 8). But his fieldwork was directed to the testing of his hypotheses, not to their generation.

Fig. 9.   Hutton's interpretation of the structure of the Isle of Arran (Craig ed. 1978, 45). The sketch, probably by John Clerk, of an east-west section of the island shows a core of granite surrounded by flanking sedimentary rocks. A comparison with fig. 5 shows how both Huttonians and Wernerians could account for many of the features of the structure of mountains (Oldroyd 1971).

Field evidence could be found for Hutton's theories of, say, veins, elevation, mountain structure, and erosion patterns. The structure of the Isle of Arran, for example, was consistent with Huttonian theory, as well as with Wernerian theory (Oldroyd 1971) (fig. 9). When George Greenough, later to become first president of the Geological Society of London, took a tour to assess the relative merits of the Huttonian and Wernerian theories, he decided that there was evidence that supported both (Rudwick 1962a).

Hutton's concept of scientific method helps illuminate an aspect of his work that has puzzled many commentators—namely, his refusal to put his theory to experimental test. In complete contrast to continental mineralogists, Hutton (1795, 251) asserted that it was not right to "judge of the great operations of the mineral kingdom, from having kindled a fire, and looked into the bottom of a little crucible." His arguments against the decisiveness of experiments were sophisticated, deeply rooted in his epistemology. He did not deny the relevance of chemistry to geology, nor was he simply self-serving, fearing that experiments would refute his theory.

What Hutton realized was that any experiment was open to a number of interpretations. Only when the scientist had clarified his conceptions was he adequately prepared to conduct and interpret experiments. When the Genevan scientists Horace Bénédict de Saussure (1740–1799) and Marc-Auguste Pictet (1752–1825) announced results that appeared to refute Hutton's heat theory, Hutton (1794c, xv) argued, "The philosophy of light and heat, which should have conducted those experiments, was in my apprehension, too deficient to derive from them the benefit which otherwise would have infallibly followed those well conducted experiments and accurate observations."

Similarly, he doubted that Lavoisier's experiments had decisively discredited the phlogiston theory. "The number of experiments, that are to be made," Hutton (1794c, xi) said, "are infinite; but, there is only one experiment which, in every case, it is proper for us to make; and it is only philosophy or general knowledge which is proper to direct the meaning of that experiment." For Hutton, experiments, like field observations, were only worth making when preceded by the formulation of theory and succeeded by careful conceptual clarification.

LATER MODIFICATIONS OF HUTTON'S THEORY OF CONSOLIDATION

John Playfair's *Illustrations of the Huttonian Theory of the Earth* (1802/1956) was the most comprehensive defense of Hutton's theory. Playfair, an eminent mathematician, a thoughtful historian and philosopher of science, but no mineralogist, reported that he was converted to the Huttonian system on methodological grounds. Its "peculiar excellence" was that it ascribed "to the phenomena of geology an order similar to that which exists in the provinces of nature with which we are best acquainted," and produced "seas and continents, not by accident, but by the operation of regular and uniform causes" (Playfair 1802/1956, 129). Playfair thought that theories of the earth should be constructed according to the same methodological principles as Newtonian mechanics. Much less sympathetic than Hutton to the method of hypothesis, he began the tradition of presenting Hutton's system as largely inductive.[2]

Playfair believed that Hutton's chief substantive innovation was his account of consolidation by heat. He stressed that this distinguished Hutton from the Wernerians, and also from earlier theorists, such as Leibniz, Buffon, and Moro, who had emphasized heat as a geological agent but who had retained an important role for water (see also Rappaport 1964; Taylor 1969).

Consequently, Playfair was somewhat embarrassed that Hutton's theory of consolidation, particularly the part dealing with compression, did not satisfy the canons of inductive reasoning. It was "a hypothesis, conformable to analogy, assumed for the purpose of explaining certain phenomena in the natural history of the earth. It rests, therefore, as to its evidence, partly on its conformity to analogy, and partly on the explanation which it affords of the phenomena alluded to" (Playfair 1802/1956, 60–61). In defense, Playfair pointed out that this was "far from assuming anything unprecedented in sound philosophy" since the theory of gravitation in physics was also a hypothesis. Nonetheless he thought Hutton's theory "would be strengthened by an agreement with the results even of such experiments as it is within our reach to make."

Playfair rather nervously accepted Hutton's principle of compression, since he thought it "reasonable to believe" that under pressure carbon dioxide would not be given off, and limestone could be melted. But he worried that this effect was not "directly deducible from any experiment yet made." Nonetheless it was "rendered very probable, from the analogy of certain chemical phenomena" (Playfair 1802/1956, 22). He described an experiment in which Black had heated "carbonate of barytes" (barium carbonate) and found that it fused *before* "fixed air" (carbon dioxide) was given off. Surely, concluded Playfair (1802/1956, 23), the "compression in the subterraneous regions" would similarly prevent the "fixed air" escaping from limestone, and render the rock fusible.

Playfair's analogy between the behavior of barytic and calcareous earth failed to convince Hutton's critics. Murray (1802, 31–32) replied that the fact "on which this argument is founded, is doubtful." He pointed out that the impure carbonate of barytes had been fused in an earthen vessel with which it reacted to form a kind of slag. This slaggy barytic earth was not, it was true, as readily fusible as the carbonate. But, Murray continued, purer barytic earth, formed by the decomposition of the nitrate, proved to be very easily fusible. Thus, "the fact cannot be admitted, that barytes, when pure, is more infusible than when combined with carbonic acid. Precisely the reverse is the case." Therefore, why should the calcareous earth be more fusible if the carbonic acid (carbon dioxide) were retained than if it escaped?

As to Hutton's claim that *all* limestones had been consolidated by heat, Playfair modified it in two significant ways. First, to such critics as Kirwan, Playfair (1802, 147) conceded that the notion that all limestones were formed from animal remains was "more general than the facts warrant." This was an important concession. Hutton had insisted they were all organic in origin in order to establish the sustained habitability of the earth. By yielding on this point, Playfair reopened the possibility that some limestones might have been deposited prior to the appearance of living beings and that rocks of similar composition might have different origins. Second, Playfair clarified a point at which Hutton had only hinted. The stratified rocks had been only *partially* fused; it was the nonstratified rocks, particularly whinstone and granite, that had been wholly fused. All mineral substances, Playfair (1802, 89) stressed, "whether stratified or unstratified, owe their consolidation to the same cause, though acting with different degrees of energy. [But] the stratified have been in general only softened or penetrated by melted matter, whereas the unstratified have been reduced into perfect fusion." Once more, he downplayed the role of fusion, as opposed to softening, in the formation of the *stratified* rocks.

Objections to the consolidation of strata by heat continued to multiply. An anonymous reviewer ("Review" 1802, 211) asked Playfair, "If fusion were the cause of consolidation, why do we find soft, incompact clay, *under* strata of limestone?" Playfair's argument that the strata had only been softened also troubled the reviewer, since "the greater part of known substances pass, by the action of heat, from a hard and solid to a fluid state, very suddenly and directly." Under the circumstances it seemed improbable that a temperature adequate to soften the rocks without fusing them could have been maintained.

Unlike Playfair, who attempted to shore up the Huttonian system with arguments, Hall revived the chemical tradition of using experiments.[3] He waited until after Hutton's death, perhaps out of respect for Hutton's skepticism about experiments, perhaps because, although Hall was an early convert to the new chemistry of Lavoisier, Hutton defended the old chemistry until his death (Donovan 1978). He examined key rocks for Huttonian theory: limestone, whinstone (basalt), and granite.

Between 1798 and 1805, Hall carried out a tricky, tedious, and occasionally dangerous series of 500 experiments in which he heated limestone under the conditions of compression that Hutton had thought would make all the difference (Hall 1804, 1812). In several of them, Hall succeeded in consolidating powdered limestone into a substance similar to crystalline marble. Surprisingly, the experiments provoked little comment, except from the indefatigable Deluc (1809), who believed that they were not applicable to the strata formed under water. Perhaps geologists were so convinced by a combination of field and laboratory evidence that calcareous rocks had been consolidated by water that they saw no reason to consider an alternative.

Hall (1798, 1805) also experimented with whinstone (basalt). The idea had occurred to him after hearing reports of glass that had accidentally escaped from a glass furnace at Leith near Edinburgh. The glass, which cooled extremely slowly, produced stony or crystalline solids, suggesting that the *rate* of cooling was important. Clearly, Réaumur and Darcet's suggestions earlier in the century that melts could produce stony substances had not been widely accepted (see chapter 3), for Hall was a competent chemist and aware of the state of knowledge in the discipline. Hall took specimens of both whinstone and lava (which he had collected on Vesuvius with the French geologist Dolomieu in 1785). He found they could be melted and cooled rapidly to form the expected product, a glass. But they could also be remelted and cooled slowly to form the product Hutton had predicted, a crystalline mass. Hall (1794) thought that granite might be such a cooled melt (see also Beddoes 1791). Unlike Hutton, Hall believed the *rate* of cooling, rather than *pressure*, was the cause of the particular character of the product in these cases.

Another Huttonian, Robert Kennedy (1805), found Hall's products to have essentially the same composition as the originals. Seeking more disinterested confirmation of this analysis, Hall sent specimens to Pictet in Geneva. Pictet reported that the fused and cooled rock was somewhat harder and denser than the original, but in all other respects very similar (Kirwan 1800). Hall (1805, 56) concluded that the arguments "against the subterraneous fusion of whinstone, derived from its stony character, seem now to be fully refuted."

The committed critics of Huttonianism denied the force of this conclusion, but began to sound very much on the defensive. Kirwan (1800, 22) said of the experiment, "The utmost effect it can produce in an unprejudiced mind is to render the origin of whins ambiguous by making them assume the appearance of a Neptunian origin, when in fact they owe it to fusion." And Murray (1802, 251) asserted, "In strict reasoning this experiment adds nothing positive to the evidence of the Huttonian system; it only removes an objection which could have been used against it." But much of the geological community was prepared to agree with Playfair (1802/1956, 80) that these experiments removed the "only remaining objection that could be urged against the igneous origin of whinstone."

Hall attempted to answer an objection to Hutton's claim that granites had cooled from a melt. Many granites contained perfect crystals of feldspar mixed with imperfect crystals of quartz. But if granite had cooled from a melt, precisely the opposite might have been expected (i.e., perfect quartz crystals mixed with imperfect feldspar crystals) since quartz was known to fuse at a higher temperature than feldspar. Pointing out that a finely ground mixture of quartz and feldspar fused at moderate temperatures, Hall suggested that molten granite was a solution of quartz in feldspar. By analogy with the freezing of brine, on cooling the two substances would be expected to separate and solidify simultaneously (Yearley 1984). Hall (1815a, 1815b, 1826) continued to work to make a version of Huttonian theory acceptable.

John Macculloch (1773–1835), who was generally sympathetic to Werner, offered field evidence for Hutton's belief that limestone could fuse. In his *Description of the Western Islands of Scotland* (1819, 48–49), Macculloch puzzled over the large areas of gneiss, traditionally one of the primitive rocks, that he found on Tiree. They contained irregular "masses of limestone." Because they were "portions of a series of parallel strata containing organic remains, . . . the masses of limestone thus found in gneiss have once been stratified, and . . . they have suffered some posterior changes by which the appearances of this disposition have been obliterated." Without quite asserting that granitic intrusions had caused these alterations, Macculloch observed that "all the varieties of gneiss are occasionally intersected by granite veins, and they are indeed almost characteristic of this rock."

HUTTON'S IMPACT ON GEOLOGY

Apart from Playfair and Hall, Hutton attracted relatively few followers, even in his native Scotland. Whether this was for institutional reasons (unlike Werner, he never had a teaching position), biographical reasons (unlike Werner, he published his theory rather late in life), or intellectual reasons is hard to determine. I am inclined to give considerable weight to the intellectual reasons. As we have seen, Hutton's theory depended heavily on his deism, his interpretation of Newtonian powers, and his adherence to a Black-Boerhaave theory of heat. It is not clear that his readers recognized all this. But if they did they might well have been unsympathetic. Most British geologists were Christians, not deists, and rejected the way Hutton's deism shaped his theory of the earth. Most continental geologists were unswayed by arguments from any theology to a theory of the earth. Hutton's interpretation of Newtonian powers, while part of an important British tradition, meant little to most geologists. And his adherence to a particular theory of heat was undercut by the rapid changes in heat theory that took place in the early nineteenth century. Add to this that Hutton's methodology was regarded with suspicion by his contemporaries, and geologists had plenty of reasons to look askance at the foundations of the system.

Turning to the particular geological claims that Hutton made, we can treat them, as he did, in three categories: his theory of erosion, his theory of consolidation, and his theory of elevation. Theories of erosion aroused relatively little interest in the years following Hutton's death and were not of signal importance in the acceptance or rejection of geological systems (Davies 1969). Theories of consolidation were of more interest, though it was a topic that concerned geologists less and less. Essentially, no one accepted Hutton's theory of the consolidation of the calcareous strata, even though it was a key innovation of his system. When someone did, as in Macculloch's case, it was as part of a theory of metamorphism rather than original consolidation. When Lyell and Scrope revived the Huttonian theory in the 1820s, they overlooked the key role that the consolidation of the calcareous rocks had for Hutton. Both repudiated "the opinion of the Huttonian geologists, that these strata are indurated by the heat transmitted to them from the interior of the globe" (Scrope 1825, 222).

Hutton had greater success with the consolidation of basalt and granite by heat. Throughout the latter half of the eighteenth century, field evidence had been accumulated for the igneous origin of basalt (Taylor 1969). Hall's experiments and the Wernerians' willingness to contemplate an originally hot ocean helped, and by the second decade of the nineteenth century most geologists, including Wernerians, agreed with Bakewell (1813, 113) "that the part of Dr. Hutton's theory which relates to the igneous origin of the basaltic rocks, is as well established as the nature of the subject will admit of" (see also Conybeare 1816; Buch 1820).

What happened was that geologists, including Wernerians, began to coalesce the theory of the consolidation of basalt (and later granite) by heat with the theory of elevation. Wernerians decided by the second decade of the nineteenth century that a theory of elevation was necessary for geology (see chapter 7). Many of them agreed with Hutton that this was linked to the intrusion of igneous rocks, though they did not accept Hutton's underlying theory of heat maintenance by a solar substance.

But the fact that Hutton's theory was entirely causal may have been equally important in shaping geologists' reactions. True, Hutton's account of the intrusion of rocks and the relationships between intruded and stratified rocks (unconformities) meant that, given two rocks in unconformable relations, we could tell their relative ages. But Hutton evinced no interested in historical geology for its own sake. Given his cyclic view of the earth and his belief in its sustained habitability by the same range of living beings, Hutton would have found the tracing of sequences of unique events advocated by historical geologists a pointless exercise. Nor did Hutton share the growing interest in fossils. As he remarked in a letter,

> being neither botanist nor zoologist in particular, I never considered the different kinds of figured bodies, found in strata, further than to distinguish betwixt animal and vegetable, sea and land objects; the mineralization of these objects being more the subject of my pursuit. [Quoted in Eyles and Eyles 1951, 323]

Nor did he adopt, let alone advocate, anything like a concept of "formation." Rather he continued to work with the mineralogists' traditional natural kinds—the earths, salts, metals and bitumens—as befitted one who was primarily interested in causal geology. In sum, Hutton made a limited but significant contribution to causal geology which his successors recognized. But the Huttonian position was distinctly a minority one in the thirty years following his death.

7
Historical Geology

FORMATIONS: THE KEY CONCEPT OF HISTORICAL GEOLOGY

Historical geology came into its own in the generation following Werner. "Formations" replaced the old commonsense mineral classes as the key concept in reconstructing the past. The implications of this shift, latent in Werner's work but never systematically explored, were traced out. The results were both constructive and destructive: constructive because "formations" proved the clue to establishing the earth's chronology; destructive because the new concept drove a wedge between historical and causal geology that was bridged only rarely in the succeeding century. In this chapter, I shall explore the foundations of historical geology, beginning by examining what early-nineteenth-century geologists meant by a "formation." Then I shall turn to the two chief tasks of the historical geologists—namely, identifying and correlating strata (including the drawing up tables and maps to show the correlations) and reconstructing past events and conditions.

Geologists unanimously attributed the concept of "formations" to Werner. Cuvier and Brongniart (1811, 8, my translation) identified the "formation" as a concept developed by "the school of Freyberg" and defined it as "a group of beds of the same or different nature, but formed at the same epoch." Classifying rocks into formations was orthogonal to classifying them into mineral classes. Geologists gradually recognized the difference. The Belgian geologist Omalius d'Halloy (1783–1875), who produced the first geognostic survey of the whole of France, and who was influenced by Cuvier and Brongniart, distinguished in a paper read in 1813, but not published until 1822:

> two principal points of view [which] seem equally to lead to the
> division of a country into physical regions determined by the
> nature of the soil: in one it is considered geologically, i.e.,
> according to the epoch of formation; in the other with respect to its

mineralogical and chemical nature. [Omalius d'Halley, (1822) translated in De la Beche 1824, 296–97].

Gradually geologists realized that, for causal theories, including genetic ones, they had to classify rocks by mineral content; for historical theories that reconstructed the earth's past, they had to classify rocks by their age. The leading German petrologist of the nineteenth century, Karl von Leonhard (1779–1862), worried that his countrymen tended to use *Gestein* (rock) and *Gebirgsart* (formation) as interchangeable terms. Alexandre Brongniart, who shared Leonhard's concern, urged that the French distinguish the study of *"roches"* (petrological units that might or might not be the same age) from the study of *"terrains"* (formations). He defined a *terrain* as a "group of several species of rocks considered to have been formed or deposited in approximately the same geognostic epoch" (Brongniart 1827, 19, my translation). Petrology and historical geology were on their way to becoming separate subdisciplines.

Werner had classified formations by the dual characters of age and mode of formation. Most of his successors dropped *mode* of formation and concentrated exclusively on age. Mode of formation, as Humboldt (1823, 1) said, "relates to the origin of things, and to an uncertain science founded on geogonic hypotheses" and had been rejected by French geologists who preferred to define a formation as an "assemblage of mineral masses so intimately connected, that it is supposed they were formed at the same epoch," a definition borrowed from "the celebrated school of Werner." Leopold von Buch (1867, 2:89–90) used an analogy to explain the same point to his fellow members of the Berlin Academy of Sciences in 1809. In numbering the houses on a street, no one considered the building material or the color of the houses. All that mattered was that the houses could be distinguished and their relative positions determined. Similarly in identifying formations, all that mattered was that they could be distinguished and their relative positions determined. Mineralogy and external characteristics were irrelevant. Brongniart (1827) responded that this was fine, as long as the geognosist recognized that formations, not rocks, corresponded to houses. The English geologist W. H. Fitton (1780–1861), who had studied with Jameson, when reviewing the third volume of the *Transactions of the Geological Society of London*, described Werner's contribution to geology as follows: "[Werner] was the first to draw the attention of geologists, explicitly, to the *order of succession* which the various natural families of rocks are found in general to present, and in having himself developed that order to a certain extent." Werner had developed the concept of formations and their determinate order such that "in the series A, B, C, D, it may happen that B or C, or both, may be occasionally wanting, and consequently D be found immediately above A; but the succession is *never* violated, nor the order inverted, by the discovery of A above the formation B, or C, or D, nor of B above those that follow it" (Fitton 1818, 71).

Some geologists, including Alexander von Humboldt, while quite committed to a geology constructed around the Wernerian succession of formations, doubted whether such an arrangement could truly be called a classification. A classification, after all, was supposed to be based on timeless essences, so that making time the essence was taxonomic lunacy. A succession, said Humboldt, could "in no respect . . . be called a classification of rocks" (Humboldt 1823, 16). Ages, or places in a sequence, were, after all, hardly the kinds of essential characters taxonomists had traditionally sought.

The use of formations as the basic unit of geology introduced a new worry for geologists, a worry about the identity and distinctness of the units being classified, or the *independence* of formations. Neither field observations nor theory readily resolved this problem. In some cases rocks were separated by sharp breaks, and thus formations did appear to be clearly distinguishable entities. Humboldt (1823, 6) remarked, "What proves the *independence of a formation*, as M. de Buch has well observed, is its immediate superposition on rocks of a different nature, which, consequently, ought to be considered as more ancient." In other cases, transitions were gradual. Furthermore the relative distinctness of formations varied from place to place, so that a sharp break found in one exposure might not be detectable in another.

Further, according to Werner's original theory, geologists had every reason to believe in gradual change, which would have had the consequence that each formation graded insensibly into the next. Cuvier postulated of worldwide "catastrophes" (1812 and 1817) which, had they occurred, rendered the independence of formations both predictable and explicable. Nonetheless, if biologists had problems with the individuation of species, and mineralogists with the individuation of minerals, geologists had far worse problems with the individuation of formations.

Yet in spite of these theoretical difficulties, the substitution of formations for rocks and the development of techniques for correlating formations and reconstructing the past was to prove enormously successful. So successful, in fact, that what had started out as one branch of the Wernerian radiation came to be accepted by geologists of all persuasions, not just Wernerians. But there were losses associated with this success.

Werner had explained the several characters of rocks and formations in terms of a single causal agency—the shifting chemical composition of the ocean in which they had been laid down. Unlike later stratigraphers, he postulated a causal connection between the order of deposition of the formations and the mineralogy of formations. In Werner's cosmogony, causal geology and historical geology still referred to many of the same entities and seemed to be complementary. But Werner's nineteenth-century successors had no explanation to offer for the relationship between the order of the formations and their mineralogy. They had to write off the question, saying that whatever relationship there might be was either mysterious or random. Not until much later did historical geology become sufficiently sophisticated that geologists could speculate plausibly about the eroded land masses that had provided the material for a given formation or about the conditions prevailing when a formation was deposited. Thus, the shift to a strictly historical geology, which produced such tremendous success in unraveling the details of the earth's chronology, was not achieved without cost. Some explanatory power was lost in the process.

THE CORRELATION OF FORMATIONS

It is to the question of correlation that we must now turn, for this dominated nineteenth-century historical geology. Historical geologists put most of their energy into determining the succession of formations in the field and then correlating successions in different parts of the world. They set aside questions about the conceptual foundations of their work, except when these were forced on them.[1] Werner had predicted that the succession of formations that he described in the *Short Classification* (1787) would be found worldwide. If geologists were to avoid the localism that had cramped the subject for so long, they had to correlate formations with some succession, correcting it where necessary. Correlation, of course, was just the geological variant of the old taxonomic problem of identification. Werner had already seen how crucial it was to geology as well as to mineralogy. Correlating formations was a slow and difficult undertaking that came to fruition only in the 1820s. It has remained a key concern until the present.

Geologists began by determining a particular succession. Since the essential character of formations was their relative age, geologists had to find some way of ascertaining whether one formation was older or younger than another. They did this by determining the succession of the formations. Because they believed that the formations had been deposited sequentially, they were confident that formations lower in the succession were older than those above them. Thus place in the succession mapped on to age of formation. Humboldt (1823, 9), using technical taxonomic terminology, declared, "The essential character of the identity of an independent formation is its relative position, or the place which it occupies in the general series of formations" (see also Omalius d'Halloy 1822, 360). Omalius d'Halloy (1822, 369, translated by De la Beche 1824) stated:

> All the other characters we employ for these determinations are but analogies drawn from the observation of places where the superposition is evident, and where it does not appear that the primitive deposition of the beds has been deranged.

Then, using Werner's succession, and assuming that the characteristics of the formations he had described did not vary from place to place, geologists attempted to correlate the two successions. They used whatever characteristics they could—mineralogy, texture, height above sea level, dip, and fossil content.

At first, mineralogy was the most important indicator. Cuvier and Brongniart, for example, used it extensively in their survey of the area around Paris. They identified the distinctive white chalk formation as the uppermost of Werner's *Flötz* (stratified) formations. Then they determined the overlying succession. Since the formations were only gently inclined and regularly superimposed, this was relatively easy. They traced the outcrop of each formation by assuming that exposures with the same mineral composition belonged to the same formation.

Thomas Webster, secretary of the Geological Society of London, was sufficiently impressed by Cuvier and Brongniart's survey that he tried to use mineralogy to determine the succession of the Isle of Wight (Webster 1814, 161). Like them, he began with the chalk. But he quickly discovered that even determining the succession was not so easy, particularly where the formations were contorted. The chalk strata that formed the central ridge of the island were quite strongly folded and vertical, not horizontal. On each side of the chalk were other steeply inclined formations, making it difficult to work out the order of the formations. Webster had to assume that the chalk had originally been deposited as a horizontal layer and that it was part of the same formation found in France and other parts of southern England. On the basis of similar mineralogy, he decided that all the beds that occurred below the chalk on the English mainland were present on one side of the chalk in the Isle of Wight. Since, according to the "law of the Wernerian school the order of the beds is never inverted," the strata on the other side must have been deposited later than, and above, the chalk (Webster 1814, 165). Webster's hypothesis was supported by the fact that the beds furthest from the chalk, the Alum Bay beds, had the same lithology as the London clay, one of the most obvious of the beds overlying the chalk on the mainland which had already been described by James Parkinson (1811) (fig. 10).

Sometimes Webster could not find rocks that exactly corresponded to those in the French succession. He (1816, 209) had to argue that the London clay, for example, was equivalent to a *mixture* of two of the French beds—the plastic clay and the *calcaire grossier*:

> If we could suppose a blending or mixture between the French plastic clay, which is blackish and contains organic bodies, and the lower beds of the calcaire grossier with its green earth and fossils, we should have a compound agreeing sufficiently near with our London clay under all its varieties; with this difference, that that of the French basin would have a greater proportion of calcareous, and ours of argillaceous matter.

Sometimes no equivalent whatsoever could be found; the upper beds of the *calcaire grossier* did not appear to exist in England.

Difficulties like those Webster experienced made geologists uneasy about relying too heavily on mineralogy for correlation. Werner himself had acknowledged, in his concept of "formation suites," that a rock with a given mineralogy could appear at more than one place in the succession. Others also realized that "nature has been able to produce similar rocks at different epochs, and that consequently mineralogical characters are insufficient to determine geological divisions" (Omalius d'Halloy 1822, translated by De la Beche, 1824, 299).

Consequently, historical geologists looked for other characters of formations to supplement or supplant mineralogy. One that was tried briefly was height since, according to Werner, the younger rocks outcropped at lower altitudes. In an effort to ascertain heights accurately, Humboldt and von Buch carried barometers on their travels. Between them, they accumulated the first accurate compi-

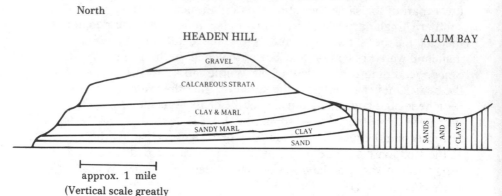

North-South Section across the Isle of Wright

North

HEADEN HILL   ALUM BAY

GRAVEL

CALCAREOUS STRATA

CLAY & MARL

SANDY MARL   CLAY

SAND

SANDS   AND   CLAYS

approx. 1 mile
(Vertical scale greatly
exaggerated)

The succession of horizontal
formations in the Paris Basin

Name of   Predominant
formation   lithology

Alluvium   Gravels ·

Upper freshwater   Marls and millstones · · · · · · · · · · · · · · · · · · ·

Upper marine   ⎧ Sandstone
                                     ⎪ Sand and sandstone ·
                                     ⎨   without shells
                                     ⎩ Marls

Lower freshwater   Gypsum and marls ·

Lower marine   ⎧ Course limestone and sandstone
                                     ⎨ (Calcaire grossier) ·
                                     ⎩ Clays (plastic clay)

Chalk   Chalk ·

Fig. 10. Webster's correlation of the successions in the Paris Basin and the Isle of Wight. Compiled from the text and illustrations in Webster 1814. To make this correlation, Webster had to depend on fossils as well as lithology (see Table 4) and to postulate a dramatic episode of elevation between the deposition of the lower Marine Formation and the Lower Freshwater Formation that heaved the Chalk and the Alum Bay beds into vertical positions.

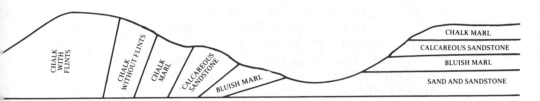

South

CHALK HILLS

ST. CATHERINE'S DOWN

CHALK WITH FLINTS

CHALK WITHOUT FLINTS

CHALK MARL

CALCAREOUS SANDSTONE

BLUISH MARL

CHALK MARL

CALCAREOUS SANDSTONE

BLUISH MARL

SAND AND SANDSTONE

Webster's correlations with strata in the
Isle of Wight

· · · · · · · · · · · · · · · · Gravel of Headen Hill

· · · · · · · · · · · · · · · · Calcareous strata of Headen Hill

· · · · · · · · · · · · · · · · Clay and marl of Headen Hill

· · · · · · · · · · · · · · · · Sandy marl of Headen Hill (no equivalent for gypsum)

· · · · · · · · · · · · · · · · Alum Bay beds (no equivalent for calcaire grossier)

· · · · · · · · · · · · · · · · Chalk with flints, without flints and chalk marl lation of the heights of some major mountain ranges (Cannon 1978). Another characteristic that was tried was dip, since many geologists believed that older rocks always dipped more steeply than younger ones (Omalius d'Halloy 1822). But geologists discovered that neither of these characteristics correlated with others, particularly mineralogy and fossil content; therefore, they rejected them as unreliable.

Over the first twenty years of the nineteenth century, geologists came to rely more and more heavily on fossils for correlation. Where Werner had listed them as an indicator of the major classes of formation, later geologists found fossils sufficiently reliable that they elevated them to the *essential characters* of individual formations. Cuvier and Brongniart (1811, 10–11, my italics) explicitly stated, in the case of the Chalk, "The exterior and small (en petit) [form and mineralogical] characters are the least useful for geological discrimination." They continued "What *essentially characterizes* this formation is the fossils it contains, fossils completely different, not only the species, but often the genera from those contained in the calcaire grossier." Their choice of the words "essentially characterize" must have been deliberate. Both were more than sensitive to taxonomic niceties. Cuvier was the leading vertebrate systematist in France; Brongniart had written on reptile classification and mineral taxonomy (1807). Their statement was a public declaration that they considered fossils—not mineralogy, for example—the essential character of at least this formation, the character that was determined by, and hence revealed, its age.

To understand the novel, indeed outrageous, nature of this assertion, we need to reconsider the principles of taxonomy. Classification of formations, as of any other set of objects, had to be based on essential characters. Mineralogists had never thought fossils essential to rocks. Rather, fossils were "accidental" in the traditional Aristotelian sense of being characters that could be present or absent without changing the character of the rock. Just as man remains man regardless of the alteration of the accidental character of his hair color, so, it was thought, limestone remains limestone regardless of the presence or absence of fossils. Eighteenth-century authors commonly referred to organic fossils as "accidental" (Bertrand 1763; Launay 1779).[2] Many mineralogists believed that fossils had been randomly deposited in the strata. John Harris (1704), for example, described "Adventitious" fossils as "reposited in the Earth by the Universal Deluge." If the dumping were that random, then there was no reason to expect fossils to be reliable indicators of particular rocks.

The analogy, drawn by Bergman among others, between the geognosist's use of fossils and the antiquary's use of medals and coins should not mislead us into thinking that fossils had been used for *correlating* rocks. In the eighteenth century, Roman coins and medals (the Greek empire had not yet been opened up to exploration) were not generally associated with specific levels in ruins or remains.[3] The location at which they were found (if anyone paid any attention to it at all) was usually, and with good reason, assumed to be largely the result of chance. And as coins and medals did not carry dates, their age, at best, could be estimated by independent evidence about the reign of the emperor or the time of the battle that was depicted. Mineralogists did use fossils, but they used them to *reconstruct* a rather generalized picture of the past, particularly of the former extent of the ocean, just as antiquaries used the portraits and symbols stamped on medals and coins to recreate a picture of life in antiquity.

Even more important, as long as naturalists believed that all species had been created at one time and that none of them had gone extinct, they had reason to suppose that any fossil might be found in any rock (or at least any rock formed subsequent to the creation of life). Of course, differences in enviornment might lead to the deposition of different fossils, but if that were the case, fossils did not indicate rocks of the same *age* of formation, but only of the same *mode* of formation.

Consequently, the naturalists and the geologists who first advocated the theory that species had gone extinct significantly overlap those who first used fossils for correlation. Fossils that had gone extinct had limited time ranges; therefore, they could be used to correlate fossiliferous formations. Even nonfossiliferous formations could be dated provided they were interbedded with fossiliferous ones.

Parenthetically, it should be noted that most naturalists believed that species had gone extinct as a result of environmental changes; hence, they tended to use fossils to reconstruct past conditions as well as to correlate formations. That was to lead to problems to which we shall return later in the chapter.

Blumenbach, who had close ties with the Wernerians, was responsible for the upsurge of interest in fossils (LeNoir 1982). In his widely distributed and readable handbook, the *Beyträge zur Naturgeschichte*, Blumenbach, unlike earlier eighteenth-century thinkers, rejected the hypothesis that all living creatures were linked in a "Great Chain of Being." He thought that altogether "too much

has been made of the matter" of the principle of plenitude. Plenitude
and the Great Chain of Being were both artifacts of taxonomy, not
representations of the real world. He scoffed at those, such as
Bonnet, Voltaire, Rousseau, Linnaeus, and Haller, who subscribed
to these doctrines:

> Nature . . . . will not go to pieces even if one species of creature
> dies out, or another is newly created,—and it is more than merely
> probable, that both cases have happened before now,—and all this
> without the slightest danger to order, either in the physical or the
> moral world, or for religion in general. [Blumenbach 1806–11/1865,
> 281–82).

Nor would he allow, as some had suggested, that fossils filled the
gaps in the Great Chain of Being. Instead, Blumenbach (1806–11/
1865, 283) argued,

> Every paving stone in Göttingen is a proof that species, or rather
> whole genera, of creatures must have disappeared. Our limestone
> swarms likewise with numerous kinds of lapidified [*versteinerun-
> gen*] marine creatures, among which, as far as I know, there is only
> one single species [of *Terebratula*] that so much resembles any one
> of the present kinds, that it may be considered the original of it.

He boldly claimed, "Not only one or more species, but a whole
organized preadamite creation has disappeared from the face of our
planet." Furthermore, not just one, but a whole sequence of such
extinctions had occurred. He cited belemites, ammonites, and large
groups of land mammals as examples of different kinds of animals
that had become extinct. He rejected out of hand Voigt's argument
that the Thuringian strata showed that fossilized land mammals had
been floated out to sea on rivers rather than being wiped out by
catastrophes. He was unwilling to allow that a single catastrophe—
the Flood—had brought about all the various extinctions at one
time. The doctrine was unnecessary theologically, besides being
hard to reconcile with animal habits. "The pilgrimage which the
sloth (an animal which takes a whole hour in crawling six feet) must
in that case have performed from Ararat to South America, is
always a little incomprehensible" (Blumenbach 1806–11/1865,
285–86). He concluded that a whole series of events had caused
successive extinctions.

Blumenbach, who admired Kant, accepted the theory of ideal
types. These types were generated by the "formative force"
(*Bildungstrieb*), a force that could never be actually observed, but
that was closely linked to the material basis of the organism, and

whose character could be deduced from its external effects. Blumenbach claimed to have developed this theory from a consideration of the fact that polyps can regenerate lost limbs, but that these are always smaller than the originals. As the *Bildungstrieb* was reduced, so too would the structures it produced diminish in size. In particular, the largest forms of a specific type might flourish in a favorable habitat, such as the tropics, while smaller organisms with essentially the same structure would be found in cooler zones. Similarly, substantially altering the material constituents on which the *Bildungstrieb* was dependent for its action might produce radically different forms. Thus, changes in the environment could produce degenerations from, or even total loss of the original ideal types. Species within a genus were quite possibly degenerations, according to Blumenbach.

Blumenbach believed that the living beings of each successive creation would be similar, but not identical, to those of other epochs. Each of the types would appear in a somewhat modified form as different environmental conditions worked on the *Bildungstrieb*.

> The Creator took care to allow in general powers of nature to bring forth the new organic kingdoms, similiar to those . . . in the primitive world. Only the formative forces having to deal with materials, which must of course have been much changed by such a general revolution, was compelled to take a direction differing more or less from the old one in the production of new species . . . so that the formative power of nature in these remodellings partly reproduces again creatures of a similar type to those of the old world, which however in by far the greater number of instances have put on forms more applicable to others in the new order of things. [Blumenbach, 1806–11/1865, 287]

These new creations were not degenerations of types. To prove this, Blumenbach pointed out, among other things, that the direction of twisting of the coils of the sea shell *Murex* went in opposite ways in different epochs, which could scarcely result from degeneration. Blumenbach urged that fossils be studied with greater care in order to reconstruct "the history of the changes of the earth's surface."

Ernst von Schlotheim and Cuvier, two of the strongest advocates of the twin doctrines of extinction and the correlation of formations by fossils, were closely connected to Blumenbach. Schlotheim studied with Blumenbach, and Cuvier with Kielmayr, who was a pupil of Blumenbach. This does not prove influence, of course, but it is at least highly suggestive.

Schlotheim began to examine fossils in the early years of the nineteenth century. Significantly, he published his results in the first volume of the *Magazin für die gesammte Mineralogie, Geognosie und Mineralogische Erdbeschreibung*, a journal started by friend and fellow student of Blumenbach, von Hoff (Schlotheim 1801a, 1801b). By 1804, Schlotheim's investigation of the plants of the bituminous schists of Thuringia (plants that would now be assigned to the lower Permian) had convinced him that they had all become extinct. He advocated their use for the correlation of formations and, as we shall see, for reconstructing the earth's history.

Cuvier encountered Kielmayr when he entered the Karlschule in Stuttgart in 1784, a school that combined a strong emphasis on moral education with a thorough training in other subjects of a more practical and intellectual character. Kielmayr himself did little paleontological work, but he wrote one of the founding assays of German *Naturphilosophie, Über die Verhältnisse der organischen Kräfte* (1793). In this, he argued that Blumenbach had gone further than warranted in suggesting a strict parallelism between earth history, ontogeny, and paleontological development (Coleman 1973). Even so, we may speculate that Cuvier was introduced to Blumenbach's ideas by Kielmayr (though it should be added that, later in life, Cuvier expressed deep reservations about *Naturphilosophie*).

From the start, Cuvier studied fossils as well as extant species. Soon after embarking on his astonishingly successful career as scientist, administrator, and educator in Paris in 1795, he began producing a series of memoirs on different fossil vertebrates, which were collected and published in 1812 (Rudwick 1972). The area round Paris was extremely rich in fossils, compared with the areas previously studied by the Wernerians. Cuvier's own experience of the Jura mountains, the Stuttgart area, the Maestrich quarries, parts of Italy, Holland, and northern Germany had failed to turn up any strata as fossiliferous (Coleman 1964). Moreover, the massive building program of the Napoleonic Empire was the occasion for extensive quarrying around Montmartre, where some of the best specimens were to be found.

Blumenbach was happy to base his arguments concerning the extinction of many species and genera on simply looking at the fossil contents of older formations. Cuvier offered a more rigorous demonstration, designed to counter the objection that naturalists might yet find living specimens of species supposedly extinct. Cuvier suggested that the large fossil quadrupeds offered the best evidence

for extinction. The chance that any of these still remained undiscovered at present was very small—something that could not be said
of marine mollusks, for example.[4] On anatomical grounds, Cuvier
decided that the elephant or "mammoth" found in Siberia was a
species distinct from either of the two living species of elephant, the
Indian and African (Cuvier 1799). (Blumenbach reached the same
conclusion independently, suggesting that the modern species had
degenerated from some original stock.) The fossil quadrupeds of the
Paris region, Cuvier said, similarly belonged to species that were
now extinct.

The first stage in using fossils for correlating formations, as in
using mineralogy for correlating formations, was to use the succession to determine the sequence of fossils. Thus, Cuvier and
Brongniart used the succession around Paris to determine the
sequence in which fossils appeared. That done, the fossils in an
unknown formation could be used to correlate it with a known
formation. Webster correlated the succession in the Isle of Wight
with that in the Pasin Basin by comparing the sequence of fossils
Cuvier and Brongniart had found with those he discovered. Rather
to their surprise, geologists found that, using these techniques, they
could trace much finer divisions of the succession over much greater
distances than they had anticipated. Cuvier and Brongniart traced
the thin individual beds forming the *calcaire grossier* for tens of
miles. Expectations about the precision of correlation rose as a
result.

Even so, it was to be a decade or more before, in the 1820s,
geologists could produce tables of correlation that covered much of
Europe. A necessary precondition for correlating formations by
fossils was the preparation of adequate classifications and, even
more important, identification guides to fossil fauna and flora. As we
saw in chapter 4, fossil taxonomy had been largely neglected in the
eighteenth century. Schlotheim (1813/1967, 174-5) campaigned for
the preparation of such aids. While praising "our mineralogical
writers" for pursuing "the determination of formations in which the
fossils occur," he went on to criticize them for lack of specificity:
"For the most part, we learn simply that Ammonites, Terebratulae,
Lenticulites, Turbinites, etc., occur . . . without being informed as
to the different species of these." The vagueness with which fossils
are treated in one of the few available monographs—the *Organic
Remains of a Former World* (1804–11) by the English doctor and
paleontologist James Parkinson (1755–1824)—drives home the justice of Schlotheim's remark (fig. 11). Schlotheim (1813/1967, 175)

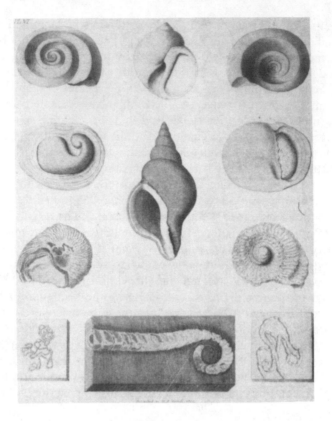

PLATE VI.

Fig. 1. A fossil shell of an unknown genus somewhat resembling *Delphinula*.

2. *Natica Canrena*, completely silicious, from Devonshire.

3. The opposite side of the fossil, Fig. 1.

4. A calcedonic cast of *Nerita conoidea*, with the containing shell.

5. The under side of Fig. 4.

6. *Murex contrarius* from Essex.

7. ⎱ The upper and under sides of a fossil of the same genus with that figured Fig. 1
8. ⎰ and 3.

9. A magnified representation of a fossil shell of the genus *Sigaretus*.

10. The same shell of its natural size.

11. A spirulite in red marble from Oeland.

12. ⎱
13. ⎰ *Vermiculitæ* in the fissile stone of Pappenheim.

Fig. 11. The identification and classification of fossils at the beginning of the nineteenth century (Parkinson 1804–1811, plate 6). Although Parkinson's engravings are clearly executed, his key shows just how undeveloped fossil taxonomy was.

argued that the paleontologist's goal should be "the accurate determination of the individual species of fossils which occur in the different formations and which appear to belong exclusively to certain of these." He further urged that these fossils should be named according to Linnaean binomial nomenclature.

Schlotheim himself classified many of the fossils that were most useful to geologists, as Jameson's appendix on Wernerian paleontology in his translation of Cuvier demonstrates (Cuvier 1817/1978). Lamarck (1744–1829) produced a taxonomy of the fossil vertebrates, which Cuvier and Brongniart used in their survey of the formations of the Paris Basin (Lamarck 1809). Since fossil vertebrates are rarely found, this classification did little to help the geologist who wanted to use fossils as a tool for correlation. James Sowerby (1812) in England and Giovanni Brocchi (1814) in Italy published classifications of marine bivalves that were of much more use. Even so the problem continued. When Lyell needed to identify the shells he had found in Sicily and Italy in the late 1820s, he had to turn to Gerard Deshayes (1797–1875) for help. Deshayes, who classified some 1,000 species of mollusks in his *Description des coquilles fossiles des environs de Paris* (1824–37), provided the taxonomic basis for the paleontological division of the Tertiary (Wilson 1972). A comparison between Deshayes's description of fossils and Parkinson's earlier effort shows how quickly paleontology developed once its foundations had been established and once there was a perceived use for it (Fig. 12).

By the 1820s, geologists were using fossils not merely to supplement other characters but to override them. The geologist who put the point most eloquently was Alexandre Brongniart. He asked himself the question

> When in two rocks, far distant from each other, the rocks themselves are of a different nature, whilst the organic remains are analogous, should we, from this difference, regard these rocks as of a different formation, or should we, from the general and *properly determined* resemblance of the organic remains, consider them of the same epoch of formation, when the order of superposition does not evidently oppose itself to this conclusion? (Brongniart 1821, 541–42. Translated by De la Beche 1824, 238).

Deciding in favor of fossils, Brongniart demonstrated that beds high in the Alps, with a height above sea level and a mineralogy that had led Wernerians to assume they were primitive, were in fact of much more recent origin. The hard black limestone outcropping at

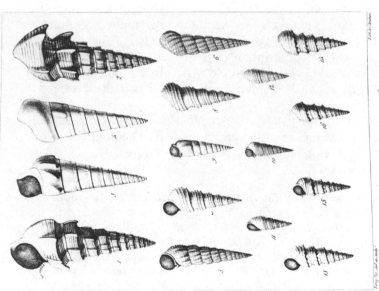

PLANCHE XII.

MÉLANIE DE CUVIER. *Melania Cuvieri.* Nob.

Fig. 1. Coquille de grandeur naturelle, vue du côté de l'ouverture.
Fig. 2. La même en dessus.

MÉLANOPSIDE DE DUFRESNE. *Melanopsis Dufresnii.* Nob.

Fig. 3. De grandeur naturelle, du côté de l'ouverture.
Fig. 4. La même en dessus.

MÉLANIE A PETITES CÔTES. *Melania costellata.* Lamk.

Fig. 5. Vue du côté de l'ouverture.
Fig. 6. Vue en dessus.
Fig. 9. Variété à côtes presque nulles.
Fig. 10. Autre variété dont les côtes ne se voient plus que vers le sommet.

MÉLANIE SOUILLÉE. *Melania inquinata.* Def.

Fig. 7. Var. du Soissonnais, vue du côté de l'ouverture.
Fig. 8. La même, vue en dessus.

MÉLANIE DEMI-PLISSÉE. *Melania semi-plicata.* Lamk.

Fig. 11. Coquille de grandeur naturelle, vue du côté de l'ouverture.
Fig. 12. La même, en dessus.

MÉLANIE SOUILLÉE. *Melania inquinata.* Def.

Fig. 15. Vue du côté de l'ouverture.
Fig. 16. Vue en dessus.
Fig. 13. Variété de la même vue du côté de l'ouverture.
Fig. 14. Vue en dessus.

Fig. 12. The identification and classification of fossils in the 1830s (Deshayes 1824–37, atlas, plate 12). Although the quality of Deshayes's engravings does not differ significantly from that of Parkinson's, the different views of each fossil and the precise identifications in the key indicate the rapid development of fossil taxonomy in the early nineteenth century.

more than 2,000 meters above sea level in the Savoy Alps contained Cretaceous fossils identical to those found in the low-lying Greensand in northern Europe (Brongniart 1821).

Earlier attempts to correlate formations by fossils had tended to be rather general. Werner, we may recall, followed common eighteenth-century practice in asserting that one of the differences between the primitive and stratified rocks was that the latter contained fossils. Blumenbach (1806–11/1865, 317) produced a threefold division based on his belief that the past had been punctuated by "revolutions." He divided fossil assemblages (and, by implication, formations) into three groups. First, there were those "whose complete similarity with still existing representatives, as well as the positions they are found in, prove that they must be comparatively the most recent." Second, there were those more or less analogous to present species, "although in climates very distant from those which contain such fossil remains," for Blumenbach had concluded that the climate had changed between the two most recent periods. Finally, there were the oldest kind, "consisting for the most part of creatures completely unknown, the records of a perfectly strange creation which has been completely destroyed." Within the oldest, entirely extinct group, there were a number of different assemblages, suggesting that a series of revolutions had occurred. Schlotheim, too, thought that fossils could be used to determine nature and extent of the revolutions that had taken place on the earth's surface (Schlotheim 1813). Omalius d'Halloy (1823) also used fossils to distinguish and correlate the stratified formations, which he named the *terrains pénéens, ammonéens, cretaceous, and mastozootic* (or formations with few fossils, ammonites, chalk fossils, and mammals, respectively).

Cuvier took Blumenbach's embryonic periodization a stage further. Believing that the fossil evidence in the Paris Basin showed extinctions to be rapid, dramatic, and simultaneous for a number of different species, he made these "revolutions"—now quite violent in character—the dividing points between his formations. In his hands, the geological use of the term "revolution" began to take on "catastrophist" overtones, rather than the simple meaning of "change" that it had had previously (Rappaport 1978). Fossils were the guide to "the revolutions of our globe" and to the "successive epochs in the formation of our earth, and a series of different and consecutive operations in reducing it to its present state" (Cuvier 1817/1978), 54–55). For Cuvier, then, formations were not arbitrary divisions of the time scale but reflections of major events in the history of the earth. The independence of formations was not a mystery but a result of dramatic upheavals in the past.

Some geologists were unhappy with this assumption of sweeping extinctions. Charles Lyell doubted that catastrophic changes had occurred. He believed that extinction, even if not proceeding completely uniformly, did not occur at occasional intervals in history with periods of stasis in between. Consequently, he developed statistical, and largely conventional, definitions of recent formations. The boundary between each of his formations was defined by the ratio of extant to extinct species, not by revolution (Rudwick 1978).

Geologists found that using fossils for correlation brought problems even as it boosted the development of historical geology. Perhaps the most intransigent was G. B. Greenough, the first president of the Geological Society of London. For Greenough (1819), the very suggestion that fossils could reveal anything about formations was anathema, even though he had studied with Blumenbach. He reasoned that, since most fossil species were extinct, inferences to age or past conditions were highly questionable. Consequently, he denied that fossils could be used to correlate rocks or to determine the conditions under which they had been deposited. As we shall see in the last part of this chapter, these were not entirely unreasonable worries, though most geologists brushed them aside.

Other problems were not so easily dismissed. One was just which fossils in a formation were to be used for correlation. Many, indeed most, fossils had ranges that extended well beyond the individual formations. When Webster wanted to correlate the strata of the Isle of Wight with those of the Paris Basin, he had to draw up quite detailed tables of a wide range of fossils using the examples that Cuvier and Brongniart had sent to the Geological Society of London (table 4). Some years later, he reflected on these problems in a letter to Brochant de Villiers, director of the French geological survey:

> Some implicitly rely upon a shell or two as furnishing certain proofs and others wish to reject almost entirely the evidence of fossil organic remains. Probably both these classes are in extremes and the proofs of identity must not be expected in any one character but in a variety of circumstances, the recognition of which demands the exercise of a judgment matured by observation and reflection. [quoted in Challinor 1961, 189]

Geologists have continued to debate how many and what kinds of fossils should be used for correlation.

Table 4. Webster's Comparison of the Fossils Found in the Upper Marine Formation of the Paris Basin with Those Found in the Isle of Wight

Fossils in the upper marine formation in the Paris Basin, according to Brongniart (1809)		Fossils in the equivalent beds in the Isle Wight, according to Webster (1814)	Linnean Names
Various beds of the upper marine marls		Names given by Lamarck	
Cytherea (bom-bees)	Ostrea Hippopus	Cerithium plicatum	
Cytherea plana	Ostrea Pseudochama	Cerithium lapidum	
Cytherea elegans	Ostrea longirostris	Cerithium mutabile	
Cytherea semisulcata	Ostrea canalis	Cerithium semicoronatum	Murices
Cerithium plicatum	Ostrea cochlearia	Cerithium cinctum	
Cerithium cinctum	Ostrea cyathella	Cerithium turritellatum	
Ampullaria patula	Ostrea spatulata	Cerithium tricarinatum	
Cardium obliquum	Ostrea linguatula	Cyclas deltoidea · · · · · ·	Venus
Nucula margaritacea	Bones and parts of Fishes	Cytherea scutellaria · · · · · ·	Venus
Patella spirorostris		Ancilla buccinoides · · · · · ·	Voluta
		Ancilla subulata · · · · · ·	Voluta

In the upper marine sand stones

Oliva mitreola	Pectunculus pulvinatus	Ampullaria spirata · · · · · · · Helices
Fusus, resembling longaevus	Crassatella compressa?	Ampullaria depressa? · · · · · ·
Cerithium cristatum	Donax retusa?	Murex reticulatus · · · · · · ·
Cerithium lamellosum	Cytherea nitidula	Bivalve apparently of the genus Erycina
Cerithium mutabile?	Cytherea laevigata	Helicina?
Solarium?	Cytherea elegans?	Murex nodularius
Melania costellata?	Corbula rugosa	Melania? They are however too much
Melania?	Ostrea flabellula	injured about the mouth to
		determine their genus with
		certainty
		Another species of Melania corresponding
		to those of Plumstead
		Natica Canrena · · · · · · · Nerita
		Ostrea, approaching to deltoidea
		Ostrea, specific characters not evident,
		but different from the last

Compiled from Webster 1814, 219–21. The identifications of both sets of fossils are very sketchy, even though Webster had the assistance of Parkinson and even though Cuvier and Brongniart were expert taxonomists. Note that Webster relies on correlating faunas rather than on type fossils and that, although there is overlap between the faunas, it is far from complete.

Moreover not all formations contain fossils. Fossils are found only in the stratified formations (and not even in all of those), so that many geologists, Humboldt, von Buch, and Jameson among them, remained uneasy about the practice of regarding fossils "as affording characters of superior importance to all others" in the identification of rock formations. "We must protest against the use [many] have made of fossil organic remains. . . . They have too often lost sight of the mineralogical relations of the rocks" (Jameson's notes, in Cuvier 1817/1978, 319–20). Similarly, in the preface of the fourth edition of his *Introduction to Geology* (1833), Bakewell inveighed against those "French conchologists" who "are endeavouring to establish the doctrine that fossil conchology, independent of the succession and stratification of rocks, is the only true basis for geology." With approval, he noted that Ami Boué, one of the founders of the Société Géologique de France, was attempting to stem this tide. It was all very well for Cuvier to trumpet the advantages of having conclusions "furnished directly for the secondary formations by the extraneous fossils" over conclusions established only by "analogy"—a vastly inferior method—for the primitive rocks (Cuvier 1817/1978, 55). Many geologists believed that conclusions based on extinct species in the secondary formations were only analogous, not direct. Even worse, Cuvier seemed to forget that the the scope of geology would contract drastically if geologists restricted their attention to fossiliferous formations.

By the 1820s the effort that had gone into surveying and correlating the formations began to pay off. Geologists published a series of comparative tables of correlation of the formations. The most ambitious of these was Humboldt's *Geognostical Essay on the Superposition of Rocks, in both Hemispheres* (1823) which was published simultaneously in French and German and translated into English within the year. A year later, De la Beche published a comparative table of the German, French, and English formations (De la Beche 1824) (fig. 13). William Buckland (1784–1856) had already published a series of tables of correlation for English and German formations (intended for his lectures at Oxford, but in fact much more widely distributed), which were incorporated into the 1818 edition of Phillips's *Outline of the Geology of England and Wales* (Rupke 1983b). He was much impressed with Humboldt's work. He equipped Lyell, his former student, with a letter of introduction and a gift of the *Reliquiae Diluvianae* (1823) and commissioned him to explain the English geological community's reception of Humboldt's correlations. In the meantime, Buckland's

A SYNOPTICAL TABLE OF EQUIVALENT FORMATIONS, BY H. T. DE LA BECHE, Esq. F.R.S. &c.

The Names having the letter B annexed are adopted by Brongniart in the succeeding "Table."—Such as are printed in Italics, are the Synonyms of those immediately preceding them.

ENGLISH.	FRENCH.	GERMAN.
ALLUVIUM & DILUVIUM.	TERRAIN DE TRANSPORT & ALLUVION (B. *Terrains de Transport* (Daubuisson.)	AUFGESCHWEMMTE-GEBIRGE.
Superior Order (Conybeare) *Tertiary Rocks*...	Terrains de Sédiment Supérieur (B)... *Terrains Tertiaires*...	Jüngstes Kalktein Gebilde & Braunkohlen Formation (Keferstein).
Upper fresh-water formation	Troisième terrain d'eau douce. (B.)	Süsswasser-Formation (Keferstein.)
Upper marine formation	Deuxième terrain marin. (B.)	Knochenführender Gyps, &c. (Kef.)
Lower fresh-water formation	Deuxième terrain d'eau douce. (B.)	Cerithenkalkstein. (Keferstein.)
Lower marine formation	Premier terrain marin. (B.)	Plastischer Thon. (Keferstein).
London Clay	*Calcaire grossier.*	
Plastic Clay	Premier terrain d'eau douce. (B.) *dep.? P?-deque.*	
SUPERMETAL ORDER. (C.) *Secondary Rocks.*	TERRAINS SECONDAIRES.	FLÖTZGEBIRGE.
Chalk. a. Upper or flinty chalk. b. Lower chalk. c. Grey chalk & chalk marl.	Craie. a. Craie blanche ou Supérieur. b. Craie tufa. c. Craie inférieure.	Kreide-Formation.
Green Sand	Craie Chloritée. *Glauconie crayeuse.* (1) *Grès & Sable vert & à kérat.*	
Iron Sand		
Oolite formation	Calcaire du Jura	Jurakalk formation.
Lias	Calcaire à Gryphées.	
Red Marl & new Red or Saliferous sandstone	Quadersandstein. Grès bizarre.	Quadersandstein. Muschelkalk. Bunter-sandstein.
Magnesian limestone	Calcaire Alpin... *a.b. &c.&c.* (Humboldt)	Alpenkalkstein. *u. Zechstein, &c.*
New Red Conglomerate	Poche rouge... (Humboldt) *Grès rouge.*	Rothe-todte-liegende.
New Red Porphyry	Porphyre du Grès Rouge. *P. phyre Secondaire.*	Porphyr gebirge. (Keferstein).
MEDIAL or CARBONIFEROUS ORDER. (Conybeare.)	Generally considered by Foreign Geologists as referrable partly to the preceding, partly to the following class.	
Coal Measures. a. Slate clay. b. Coal Grit.	Terrain Houiller. *Grès rouge.* (Humboldt). a. argile schisteuse. b. grès des Houillères... *Psammite.* (B).	Steinkohlengebirge. a. Schieferthon. b. Kohlensandstein.
Mountain or Carboniferous Limest. Old Red Sandstone	Calcaire de Trans. (O. d'Halloy, &c.) Grauwacke (Humboldt, &c.)	Grauwacke.

ENGLISH.	FRENCH.	GERMAN.
PRIMORDIAL ROCKS. *Primary Rocks* (Macculloch.) *Submedial & Inferior Orders* (Conybeare.) *Transition & Primitive Rocks.*	TERRAINS PRIMORDIAUX. (Omalius d'Halloy.) *Terrains intermédiaires & primitifs. Terrains de Transition & D^x*	Übergangsgebirge & Urgebirge.
Greywacke	Grauwacke. *Traumate* (Daubuisson) *Psammite* (B)	Grauwacke.
Greywacke Slate	Grauwacke schisteuse...	Grauwackenschiefer.
Transition Limestone *Schuerlind Limestone*	Schiste Transanique. (Daubuisson.) Calcaire de Transition. *Calcaire intermédiaire.*	Übergangskalkstein.
Alum Slate	Schiste Alumineux... *Argillite aluminieuse.*	Alaunschiefer.
Whetstone Slate	Schiste corticale. *Schiste novaculaire. Schiste à aiguiser.*	Wetzschiefer.
Flinty Slate	Schiste Siliceux... *Jaspe Schisteuse* (B)	Kieselschiefer.
Serpentine	Serpentine. *Ophiolite.* (B)	Serpentin.
Dialläge Rock	Euphotide. (Haüy).	Schillerfels.
Greenstone	Diabase... (B)	Grünstein. (Von Buch). *Grünstein.*
Greenstone-Slate	Diabase schisteuse. (B.)	Grünsteinschiefer.
Quartz Rock	Roche de Quartz. *Quartzite* (B. à l'émail)	Quarzfels.
Clay Slate	Schiste Argileux. *Phyllade* (Daubuisson.)	Thonschiefer.
Chlorite Slate		Chloritschiefer.
Talcose Slate *Steatite*	Schiste talqueux. *Stéaschiste* (B) *Stéatite*	Talkschiefer.
Hornblende-Rock	Amphibolite (Daubuisson)	Hornblendgestein.
Hornblende-Slate	Amphibolite schisteuse.	Hornblendschiefer.
Primitive Limestone *Granular Limestone*	Calcaire primitif. *Calcaire grenu.*	Urkalkstein.
Mica Slate *Compact & Granular Felspar*	Schiste micacé. *Micaschiste* (B).	Glimmerschiefer.
Whitestone? (Jameson)	Eurite. (Daubuisson) *Leptinite.* (Haüy.)	Weisstein.
Gneiss	Gneiss.	Hornfels. Gneiss.
Granite *var. Graphic Granite var. Protogine*	Granite. (Daubuisson.) *var. Graphic Granite. var. Pegmatite.* (Haüy). *var. Protogine.* (Jurine).	Granit.

ENGLISH.	FRENCH.	GERMAN.
TRAP ROCKS (Macculloch). *Overlying Rocks* (Macculloch.)	ROCHES TRAPPÉENNES.	TRAPP-GEBIRGSARTEN.
Wacke	Wacke.	Wacke.
Claystone	Variolite (?)	Thonstein.
Clinkstone	Phonolite. (Daubuisson.)	Klingstein.
Compact Felspar	Feldspath Compacte. *Petrosilex.* (B).	Dichter Feldspath.
Pitchstone	Rétinite. (B.)	Pechstein.
Hornblende Rock	Amphibolite.	Hornblendgestein.
Basalt	Basalt.	Basalt.
Dolerite *Augit Rock* (Macculloch).	Dolérite.	Dolerit.
Amygdaloid *var. of Amygdaloid*	Amygdaloïde. *Variolite*	Mandelstein. Blätterstein.
Corsean	Corneenne. *dolomite.*	
Greenstone	Diabase (B).	Grünstein.
Syenite	Syenite.	Syenit.
Porphyry	Porphyre.	Porphyr.
Clay Porphyry *Claystone Porphyry* (B)	Thonporphyr. *Argilophyre* (B)	Thonporphyr.
Clinkstone Porphyry	Pomolite...	Klingstein porphyr.
Felspar Porphyry	Porphyre eutique. *Porphyr noir.*	Feldspath porphyr.
Pitchstone Porphyry	Sigénite (B)	Pechstein porphyr.
Porphyritic Greenstone *Greenstone Porphyry.*	Mélaphyre (B) *Porphyr noir.*	Trappporphyry. Grünstein-porphyr.
Trap-Tuff.	Diabase porphyroïde.	Trap-tuff.

ENGLISH.	FRENCH.	GERMAN.
VOLCANIC ROCKS.	TERRAINS VOLCANIQUES.	VULCANISCHE-GEBIRGE.
Trachyte	Trachyte	Trachit.
Pearlstone	Perlite	Perlstein.
Basalt	Basalte	Basalt.
Lava. a. Compact lava. b. Scoriform lava. a. Porphyritic lava.	Lave. a. Lave compacte. b. Lave scoriacée. *Lave porphyroïde* (B).	Lava.
Obsidian a. Obsidian Porphyry.	Obsidienne. a. Obsidienne porphyry.	Obsidian. a. Obsidian porphyr.
Pumice	Ponce.	Bimstein.
Volcanic Conglomerate	Bréche Volcanique.	Vulcanische Breccien.
Volcanic Tufa	Tuf Volcanique.	Vulcanische Tuff.

Fig. 13. A typical table of correlation for formations from different countries. A portion of De la Beche's "Synoptical Table of Equivalent Formations" (1824).

tables, which Phillips (1818, 19–20) explained were intended to "shew the general agreement of the British strata, with the order adopted by the Wernerian school," had been widely praised on the Continent, as well as in England (Rupke 1983b). Ami Boué published a "Synoptical Table of the Formations of the Crust of the Earth" in the *Edinburgh Philosophical Journal* in 1825, Alexandre Brongniart a *Tableau de terrains qui composent l'écorce du globe* in 1829, and Amos Eaton a "Tabular view of North American rocks" in the *American Journal of Science* in 1828.

By the 1830s, then, geologists had reached agreement about the outlines of a table of succession, at least for the secondary formations, and were hammering out a common nomenclature. Gradually nonmineralogical names replaced mineralogical ones as a consequence of identifying formations by characters other than mineralogy. The new table of succession differed in many respects from Werner's early effort, as a result of geologists' extensive surveys of the stratified rocks. The succession was much longer than anyone had guessed. The discovery of the length of the geological column and the vast time spans required for its deposition created interest, but not consternation, among geologists.

Besides correlating the formations, geologists mapped the formations that cropped out at the surface.[5] This followed naturally on correlation. Geological maps were an economical and sophisticated way of representing the succession as it had been shaped by structural forces and as it was intersected by the topography. In popular terms, they were (and are) a representation of the "packed-down" theory of historical geology. Like correlation, geological mapping was often surprisingly tricky in practice. Most of Werner's identifying characters could be observed only where there were outcrops. Except in dry or mountainous regions, and in limited manmade exposures, such as quarries, mines, and road cuttings, rocks are almost invariably covered with a thick layer of soil and vegetation. In these cases, underlying formations had to be inferred from surface features, and, where possible, the inference had to be checked by examining exposures. The problem was worse with "secondary rocks" than with the "primitive rocks": "the secondary formations are less easy to observe than the primitive: more usually horizontal, it is rarer to find large vertical cuttings; and their different arrangements are not, in most cases, uniform" (Cuvier 1810, 183, my translation).

Fortunately, geologists realized surface features often co-vary with underlying formations. Arthur Aikin (1773–1854), who was a leading member of the short-lived British Mineralogical Society (1779–1806) and a chemist-mineralogist, pointed out,

> The primitive, secondary, and derivative mountains, may in general be distinguished by peculiarities in their *form*, as well as in their relative position; the primitive rocks are craggy, steep, and tending more or less to a peak, or slender pointed summit; the loftiest mountains are generally about the middles of the chain, which both commences and terminates in abrupt precipices: these, together with the insulated peaks that are continually interrupting the outline of the chain, form a very striking distinctive character. [Aikin 1797, 217]

Geologists soon used surface features to trace divisions much finer than those between the primitive and secondary classes. Even beds within formations tend to give rise to a particular type of relief and soil, and, in turn, characteristic agriculture and architecture. For example, the Jurassic oolitic limestone is fairly hard in southern England and forms the well-known Cotswold uplands. Since it is pervious, few rivers cross the outcrop, and, in the early nineteenth century, the soil was inadequate to support anything other than flocks of sheep. Houses in the region were built and roofed with the local stone. By contrast, the nearby outcrop of Oxford clay underlies well-watered valleys, containing dairy farms and villages of woodframed, thatched houses. A traveling geologist could point out the distribution of the formations without having to check exposures.

In short, mapping was a multiply theoretical activity. The formations that surveyors traced were theoretically defined entities. They were also, by and large, inaccessible to direct observation. A geological map was not a simple representation of the world but a highly complex interpretation of it.

There remained the problem of representation. When the geologist had worked out the structure and distribution of the formations in a given area, he had to find some way of conveying it to others. Geologists experimented with a wide range of representations in the late eighteenth and early nineteenth centuries (Laudan 1974; Rudwick 1976; Taylor 1985). A few attempts were made to show all three dimensions at once: some geologists made models; John Farey (1811) used block diagrams; William Smith (1815) tried shading as a representation of the third dimension. All these techniques proved unwieldy, as well as expensive. In a different vein, Humboldt (1823) worked out an ingenious pasigraphic scheme for indicating the sequence of rocks, including the formation suites. Once again, no one actually employed it.

By the 1820s, most geologists showed the vertical sequence by a series of columns or longitudinal sections. Readers were left to put together their own three-dimensional picture from these two separate kinds of information—an exercise that varied in difficulty depending on the structural complexity of the region. Sections had been common for at least a century. Sophisticated ways of representing the areal patterns of distribution took longer to develop.

In part, this reflected the undeveloped state of cartography. Most maps showed limited areas. Since the gentry usually financed by subscription, they were designed to flatter the subscribers by depicting their castles, manors, and hunting grounds. Cartographers concentrated on the wealthy and populous areas; they neglected the rugged, mountainous backwaters of greatest interest to geologists. Surveying was often careless. Before Humboldt's campaign for accurate measurement in the early nineteenth century, no heights had been measured with any precision. Scales and conventions varied from one map to another, making it next to impossible to construct a detailed large-scale map of any significant stretch of country. Conventions for representing relief were clumsy and unreliable. Hills and mountains were shown either as little "molehills" or by hachuring.

In spite of these difficulties, geologists turned enthusiastically to mapping. Werner had urged his pupils to construct maps, had personally directed some surveys, and had drawn up instructions for their coloring (see Chapter 5). Johann Karl Freiesleben, one of the first of his pupils to produce a geological map, represented the Permian and Triassic formations of Thuringia in his *Geognostischer Beitrag zur Kenntniss des Kupferschiefergebirges* (1807–15). At about the same time, William Maclure produced an essentially accurate, if very generalized, map of the major classes of formation represented on the Eastern Seaboard of the United States, which he published in the *Transactions of the American Philosophical Society* in 1809. Cuvier and Brongniart's map of the Tertiary strata of the Paris Basin and Webster's map of their equivalents in southern England appeared in 1808 and 1814, respectively.

From the start, Wernerian mapmakers wanted maps of different areas to be compatible. Webster's efforts to integrate his map of the Isle of Wight with Cuvier and Brongniart's map of the Paris Basin is a case in point. The development of compatible units, scales, nomenclature, and conventions was far from easy, even when the geologists accepted essentially the same geological theory, as Cuvier, Brongniart, and Webster did. But when the geologist had his own idiosyncratic theory and terminology, the problems of localism could become overwhelming. This is true of the two major early maps of England, the one prepared by William Smith, the other by the Geological Society of London, led by G. B. Greenough. These maps are often treated as the locus of early-nineteenth-century innovation in geological cartography (Laudan 1974). Seen in an international context, they appear provincial. But since they have played such a large role in the literature, I shall discuss at some length why I conclude this before returning to the Wernerians.

An untutored "mineral surveyor," William Smith (1769–1839) completed the first geological survey of England (Smith 1815). His map was most impressive. Measuring eight by six feet, it surpassed all earlier geognostic maps in the area covered and in the number of fine divisions traced. Smith worked quite independently of the Wernerian tradition of mineral geography, having had neither a formal nor informal education in geology (Laudan 1974). He was almost entirely unaware of work going on on the Continent. His papers contain numerous plans for further education, pathetically stopping within a page or two, but almost no books or articles on geology, British or foreign. Neither his income, nor his education and social standing, nor his life-style as a traveling mineral surveyor allowed him to enter the world of the naturalist or natural philosopher.

Yet it is perhaps not surprising that he independently developed the same techniques, and even some of the same conclusions, as the Wernerians. Like the Germans and the Swedes, Smith had strong pragmatic motivations for studying rocks. Consequently, he independently invented many of the techniques used by continental mineral geographers. He even employed the same cartographic conventions, indicating the extent of the outcrops with color washes, though his novel and graphically effective practice of representing the lower edge of each outcrop with a deeper shade was so expensive that no one else adopted it. What Smith lacked was his continental contemporaries' background in chemistry, mineralogy, and natural history. Correspondingly, his map was impressive insofar as it embodied his hard-won practical experience of British geology, but disappointing in its theoretical underpinnings and local in its division and nomenclature.

Judging by his notebooks, Smith seems to have believed that the strata (he never referred to formations) were laid down by a series of inundations sweeping England from southeast to northwest, each of them depositing a completely homogenous layer (Smith mss; Greenough 1819). In constructing his map, Smith adopted three assumptions: (1) the surface of the earth was formed of a limited number of strata that occurred in a definite order; (2) these strata had similar orientation or, in technical terms, they all had the same dip and strike; and (3) the properties of any stratum, including its surface features, remained the same throughout its extent (Smith 1815; Laudan 1976). Smith did not laboriously survey the whole ground area, but simply checked his predictions at certain key spots by employing his third assumption that the surface features are a reliable guide to the underlying stratum. Since most of the strata in southwest England do have this distribution, Smith accomplished his mapping with dispatch.

Fossils were a very time-consuming, even if revealing, way of correlating formations. Smith, who is widely believed to have used fossils in the construction of his map, frequently ignored them in favor of other more easily observable features of the strata (Laudan 1976). For fossils to be the criterion for identifying strata, they must be given precedence in cases where there is conflict between the different criteria. This was just the point at issue in the Wernerian debates about which of the congeries of characters usually found in constant association should be given priority when that association broke down. In a couple of important cases of such conflict, Smith gave priority to geographical position, mineralogy, and surface features. These identifications have often been described as "mistakes" rather than as reflecting conflicting criteria of identification (Judd 1897; Cox 1942).

For example, Smith confused the two limestones—the Magnesian Limestone and the Metalliferous or Mountain Limestone—that occurred below the Cretaceous and Jurassic strata that he knew best, even though they have quite distinct fossil faunas (fig. 14). Assuming that they must occur, like the other strata he had examined, in parallel northeast-southwest lines, he placed two discontinuous bands of color on the map that he hoped later geologists would speedily complete. Indeed, in illustrated plates for one of the accompanying volumes, Smith reversed the fossils supposedly contained in the two limestones (Cox 1942).

Similarly, Smith used mineralogy and surface features, not paleontology, when he identified outcrops in Blackdown in southwest England, the Weald in the southeast, and the North York Moors as

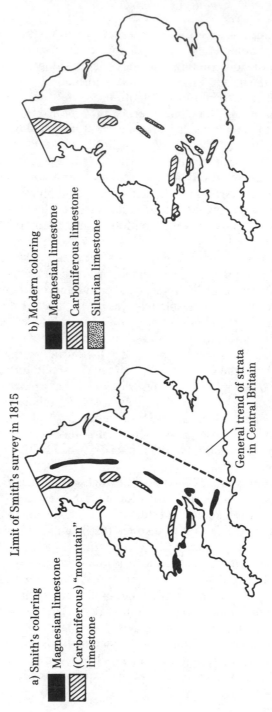

Limit of Smith's survey in 1815

a) Smith's coloring

◼ Magnesian limestone

▨ (Carboniferous) "mountain" limestone

b) Modern coloring

◼ Magnesian limestone

▨ Carboniferous limestone

⠿ Silurian limestone

General trend of strata in Central Britain

Fig. 14. Smith's identification of strata by geographical position contrasted with the identification of strata by fossils. A comparison of Smith's method of mapping the Magnesian and "Mountain" (Carboniferous) limestones (redrawn from Smith 1815a) with the present method. Smith's identification of the different limestones is based on their geographical position with respect to the generally southwest-northeast trend of most of the strata cropping out in southern England rather than their fossil content.

Chalk

Iron sand and carstone

NORTH YORK MOORS
now identified as Jurassic

BLACKDOWN
now identified as Lias

WEALD
now identified as Lower
Cretaceous or Wealden Beds

Fig. 15. Smith's identifications of the strata of the North York Moors, the Weald, and Blackdown as iron sand and carstone (redrawn from Smith 1815a), based on surface features rather than fossil content. Today the strata underlying these three areas are assigned to different parts of the geological column.

belonging to the same stratum. All three are covered by heathy moorland, and underlaid by poor sandy soil, but their fossil content differs markedly, and they are now assigned to different parts of the geological column (Laudan 1976) (fig. 15).

In 1831, Roderick Murchison succeeded in getting Smith dubbed the "father of English geology" (Geikie 1905/1962; see also Cox

1942; J. Eyles 1969; Torrens 1978; Rudwick 1985). Smith's success in producing a map of the British strata does not, in itself, establish that he was responsible for the surge of interest in historical geology in Britain. Most British geologists learned their mapping from the continental tradition, not Smith. True, Smith did demonstrate his results at agricultural shows from the turn of the century on, and he did teach a few other mineral surveyors, such as John Farey and Benjamin Bevan. Nonetheless, he had few contacts with the leisured amateurs who made up the bulk of the English geological community, and his publications did not appear until after 1815, by which time Wernerian methods had become standard practice among English geologists as the early *Transactions of the Geological Society of London* show (Porter 1977). Smith can be pitied for his isolation and admired for his determination in the face of it, but it is hard to imagine that the development of geology would have been much different if he had never published.

The members of the Geological Society of London prepared their own map of England, and it was published in 1820, just five years after Smith's. It was prepared in conformity with Greenough's urgings that the society adopt a strict inductivism that, while giving a characteristic stamp to their activities, would be unassailably scientific. According to Greenough, the society's hallmark should be enumerative induction—that is, the collection of facts unsullied by any theoretical presuppositions, from which theory might, in the fullness of time, emerge (Hull 1973; R. Laudan 1977b). The method had the twin virtues of being regarded as archetypically, if conservatively, scientific and of being employable even by those completely ignorant of the science in question. The society could actively recruit London and provincial members, even though most were ill-versed in mineralogy and geology, to gain strength from numbers.

In April 1809, the society established a Committee on Maps, charged to undertake the society's major cooperative project, the preparation of a geological map. Every provincial member was to be urged to submit details of local rocks and strata, collected without any theoretical bias. The London members were to collate this information and enter it on the map. Gradually a "system" would emerge, and the outlines of English geology would become clear, vindicating the society's existence.

Although nominally a collective activity, the map was largely financed and prepared by Greenough. As his diary and journals and the mass of correspondence he preserved testify, he approached the task with as thoroughgoing a commitment to enumerative induction as can be found in the history of science. From provincial amateur members of the society, from his personal friends, from books and journals, and from annual summer field trips, Greenough systematically collected information about the rocks of Britain, entering it in a series of hardbound notebooks (Greenough mss., University College, London [notebooks] and Cambridge University [correspondence]). The quality varied. On geological tours, Greenough "catechised . . . Innkeepers Waiters Ostlers Postilions Wellsinkers Masons Gamekeepers, Mole Catchers, . . . one after another and if their acts [varied] confronted and cross examined [them]" (Greenough to "Mrs. S.," 22 August 1814, Greenough mss., Cambridge). Many contemporaries were appalled. "This coxcomb's reign must soon be over," exclaimed one. "The day is gone when a man could pass for a geologist in consequence of having rattled over 2 or 3000 miles in a postchaise and noted down the answers of Paviors, road menders, brick makers, and Limburners" (Underwood to Thomas Webster, quoted in Challinor 1963, 54).

Although heavily indebted to Smith's map, Greenough improved on it in certain respects. Yet his resolute refusal to confront the conceptual foundations of the task in which he was engaged, rendered the map less useful than it might have been. Brochant de Villiers, by now head of the French geological survey, commented that, while cartographically the map was "un chef d'oeuvre d'exécution," it was of purely provincial interest since no attempt was made at correlation. "His map such as it is will be very useful for the English or for mineralogists who will travel in England, but *there is nothing in it for Science*" (letter to Webster, 1820, quoted in Challinor 1961, 184–85, my translation). To be a true contribution to science, he suggested, there had to be an analysis of the relations between the different rocks and a correlation of the English rocks with their continental counterparts.

Meanwhile, Wernerians continued to produce maps that attained some degree of compatability between different regions. Omalius d'Halloy published a map of France and Belgium and the surrounding parts of Germany and Switzerland in 1822. Brochant de Villiers campaigned for a map of France and, with Armand Dufrénoy and Élie de Beaumont, began his survey in 1825. The map was eventually published in 1841, after his death. In 1826, Leopold von Buch anonymously published a geological map of Germany in forty-two sheets, which covered a larger area of Europe in detail than any other at the time of its publication. In the same year, John Mac-

culloch commenced a geological survey of Scotland, and the map
that resulted was published posthumously in 1836. And in 1835, the
Geological Survey of Great Britain was established under De la
Beche's directorship.

Mapping had become the stock-in-trade of the geologist. A map,
said von Buch, dean of Wernerian geologists,

> is always a decisive criterion of they who aspire to the rank of
> geologists,—every one who has not compiled a map, wants the
> necessary talent of combination. The spirited Darwin, with all his
> remarkable vivacity of mind, is for me no Geologist, only an able
> history maker of what nature as he beleaves [sic], has done, and
> what never she did. This man could never make a tolerable
> geological *map*. [von Buch to Murchison, 20 April 1846,
> Murchison mss., Geological Society of London][6]

RECONSTRUCTING THE EARTH'S HISTORY

In the early nineteenth century, one particular method of reconstructing the earth's history—using the testimony of the rocks
themselves—came to predominate over other methods that had
been employed earlier. The testimony of the ancients had, by now,
been shown to be a sketchy and incomplete source, if, indeed, it
could be relied on at all. Of course, some biblical geologists lingered
on, particularly in Britain. But the more reputable of these, such as
William Buckland, laid the onus of their reconstruction on the rocks,
not on the Bible. Genetic cosmogonies employing the laws of the
natural sciences were regarded as speculative and unscientific, at
least if used as the primary source for reconstruction; only if they
were consistent with the testimony of the rocks were they acceptable. Laboratory analogies were used less frequently as geologists
became concerned about whether processes in the laboratory and
processes in the natural world were similar enough to bear the
burden of the method.

That left inferences from the rocks themselves. Early Wernerians
had reconstructed the conditions under which rocks and minerals
had consolidated by drawing inferences from their "glassy,"
"stony," or "crystalline" texture. With the new enthusiasm for
correlating formations by fossils, and the consequent emphasis on
the stratified rocks, this mode of inference became unimportant, for
all the stratified rocks were "stony." "Glassy" and "crystalline"
rocks were of little interest to the historical geologist (though they
continued to concern the causal theorist, as we shall see).

Inference from the texture of the rocks was replaced by inference from the structure of the formations and from fossils. A whole series of inferences from the structure of formations led geologists to abandon Werner's hypothesis of a gradually shrinking ocean. They substituted the theory that formations had been elevated above the ocean by some force (probably a force within the earth, since astronomical causes were viewed with disfavor), a force, furthermore, that acted intermittently, not regularly. Among these lines of evidence were the presence of the young formations on the very top of mountain chains (Brongniart 1821); the sharp breaks often found between formations (Cuvier 1817/1978); the faults and folds in young, as well as old, formations (Webster 1814, 201; Conybeare and Phillips 1822, 111–12); and the "denudations," or areas where the upper portions of the strata seemed inexplicably to be missing (Richardson 1808; Farey 1811; Webster 1814). But such evidence was of little use in determining the nature, as opposed to the timing, of these events, or indeed in determining what the intermittent elevating force might have been. For that, some kind of causal theory was required, a question we shall address in chapter 8.

Fossils became ever more important to geologists who wished to reconstruct the earth's past. Their use was not new. Eighteenth-century geologists, assuming that species found in older formations were identical to living species, had used them to infer former positions of land and sea and former climates. In the nineteenth century, the reconstruction of former climates became a major concern.

Yet ironically the use of fossils for reconstructing the past became more problematic in the early nineteenth century as a result of the very discovery that had encouraged the use of fossils for correlation—the discovery of extinction. The dilemma was the following. The fossil species in any given outcrop were determined by a combination of the age of the formation and the environmental conditions when it was deposited. Geologists had to make sure that the fossils they were using to date formations were present as a result of the period, not the environment in which the fossils had flourished, and that the fossils they were using to infer past conditions were present as a result of the environment, not the period in which they had flourished. It took a while for the problem to sink in.[7]

Schlotheim (1804) claimed that Thuringia must once have had a warmer climate since its extinct plant species resembled those now found in the tropics. Cuvier and Brongniart (1808) relied on evidence from fossil mollusk species of the same genera as extant mollusk species to argue that the Tertiary formations of the Paris Basin had been laid down in alternating freshwater and marine conditions. In short, they treated a direct inference as straightforward and did not argue that function could be inferrred from structure, and then environmental conditions from function (Rudwick 1964).

This practice was soon to be called into question. At issue was whether the methodological warrant for the inference to past conditions was inductive or analogical—categories that for most geologists were virtually equivalent to legitimate or illegitimate. It was Charles Lyell, rather than a Wernerian, who gave the problem the most sustained analysis. In 1826, he rehearsed the standard arguments for climatic change from fossil evidence. Fossils of *extinct* species found in the strata of Europe and North America bore a striking resemblance to *extant* species living in tropical regions. All the evidence—the remains of elephantlike mammals on the steppes of Russia, the tigers and hippopotami of the Paris Basin, and the tropical plant remains in the widespread coal deposits of the Northern Hemisphere—suggested that "the former temperature of the northern hemisphere was much higher than it is at present" (Lyell 1826, 525-27). Since all the species involved were extinct, "we reason only from analogy when we draw this conclusion. . . . We ought therefore to require a great accumulation of evidence, together with perfect harmony in the proofs." And there were problems with the analogy. For example, some contemporary genera are represented by at least one species in almost every climatic region; we find the musk ox in the arctic, the common ox in the temperate, and the buffalo in the equatorial regions. Nonetheless, Lyell believed that, provided that the evidence was drawn from a large number of species, "the argument from analogy is unimpaired" and the conclusion that climate had changed was well-founded.

Even so, Lyell continued to worry about the soundness of these analogical arguments. Discussing the related problem of how to decide whether a given formation had been laid down in fresh- or saltwater—a problem whose resolution also depended on reconstructing the habitat of the fossil species in the formation—Lyell (1827, 445) decided that the method of analogy would work only if "the aquatic plants and animals found in a fossil state should belong to genera now peculiar to lakes or rivers,—or if with such be associated some belonging to extinct genera, these must fall under some great family characterized by *an organization not peculiar to marine animals*."

Then, in 1829, a Scottish biologist John Fleming (1785–1857), a pupil of the Wernerian Robert Jameson, posed a fundamental challenge to the use of fossils for reconstructing the past. Fleming reminded his readers that when fossil mammals such as the Siberian rhinoceros were first found, they were believed to be members of extant species; thus, the "conclusion seemed to be warranted, by *an extended induction*, that our region once enjoyed a tropical climate" (Fleming 1829a, my italics, 278-79). But Cuvier's demonstration that species had become extinct destroyed this warrant. Now "the whole argument in favour of a change of climate seems to depend on the value of *analogy*, as an instrument of research." Fleming asked, "Supposing ourselves acquainted with the habits and distribution of one species of a genus, can we predicate, with any degree of safety, concerning the habits and distributions of the other species with which it is generically connected?"

Fleming's answer was an unequivocal "No." Analogies between the species within a genus, at least as far as habitat was concerned, were slight, and the method of analogy was a hopelessly inadequate instrument for the reconstruction of past climatic conditions. The method of induction had been reliable; the method of analogy was not to be trusted. Two species with the same structure, such as the polecat and the otter, might have very different habits. Two species with the same external appearance, such as the common shrew and the water shrew, might have very different habits. Two species, similar in form and in structure, such as certain species of oxen, might live in quite different environments. The mammoths of Siberia, as Cuvier had shown, could well have been suited to a cold climate, despite their superficial similarity to elephants. Thus, while "in reference to the *individuals of a species*, our knowledge of the ordinary habits and distribution of a few of these qualifies us for judging respecting the remainder," the same was not true of different species belonging to the same genus. Here, "the truths of zoology forbid us to reason concerning the *species of a genus* in the same manner as we do with the *individuals of species*" (Fleming 1829a, 283). Arguments from the form of extinct species to the nature of past climate (or any other aspect of their habitat) were unreliable. In the absence of any reliable method of determining past climate, philosophic caution dictated that geologists should assume that the earth's temperature had remained constant.

W. D. Conybeare (1787–1857) was quick to reply to this threat. A successful argument that it was "impossible to reason from generic affinities as to the geographical distribution of particular species" undercut the (supposed) laws connecting the form of a species and its habitat and distribution, which were "obviously the only basis on which philosophical reasoning can be built" (Conybeare 1829, 143). Conybeare reasserted that the "cumulative character" of the evidence was crucial. "Each of these analogies [between extinct and extant species], taken separately, must surely, unless it can be neutralized by some countervailing argument, be allowed to constitute a probability"; and since "*all* the analogies *invariably* lean one way," Conybeare dismissed Fleming's attack.[8] Fleming replied that, Conybeare's claim to logical skills notwithstanding, Conybeare was unable to appreciate the force of his criticisms.

Lyell realized that, to combat Fleming's skepticism, he had to give a more adequate methodological foundation for the use of fossils to reconstruct earlier environments. In the *Principles of Geology* (1830–33, 1:93), Lyell took up the gauntlet thrown down by Fleming and argued that earlier climatic conditions could be reliably reconstructed. He gave "direct" proofs, better founded than "reasoning by analogy," that the climate of the Northern Hemisphere had become steadily cooler in the period since the appearance of the earliest fossiliferous strata known at the time—since what we would now call the base of the Carboniferous.

Since the interpretation of the data that putatively supported the thesis of climatic change had been so much disputed, Lyell began by considering the relationship between extant species and their environments. In his opinion, this was one of the few areas relevant to geology where the investigation of the present, observable state of affairs had not been neglected. For example, Humboldt had investigated the relationship between climate and relief in order to understand the effects of environmental conditions on life. The exploration of the Southern Hemisphere had brought many previously unknown species to the attention of naturalists. The experienced scientist could recognize those specimens that had been "collected from latitudes within, and those which were brought from without the tropics" even without knowing their provenance (Lyell 1830–33, 1:93). Thus, Lyell, despite Fleming's arguments, was satisfied that in most cases the organization of an extant species is a reliable guide to the environment in which it lives. But with extant species, this conclusion could be checked by observation. What were geologists to do as they went further into the past and found more and more extinct species?

Lyell answered that many past climates could be reconstructed using extant, *not* extinct, species and inductive *not* analogical arguments. Therefore, Fleming's caustic criticism of analogical methods was irrelevant. Lyell shifted attention from the fossil mammals of extinct species, on which geologists had formerly tended to rely and which Cuvier had shown to be so misleading, to fossil mollusks of extant species. He broke down the abrupt division between present and past on which Fleming's attack depended and which was canonized in much of the geological literature. Lyell, who did not believe that a recent and widespread deluge had wiped out most previous forms of life, realized much more clearly than many of his contemporaries that not all species found in the past (i.e., not all fossilized species) were extinct species. This was particularly the case with marine mollusks, which were much less susceptible to extinction resulting from environmental changes than mammals. In 1828 and 1829 in Italy and Sicily, Lyell had found many fossilized species that were still extant in some part of the world. Lyell (1830–33, 1:93–94) concluded, "It is not in strata, where the organic remains belong to extinct species, but where living species abound in a fossil state, that a theory of climate can be subjected to the experimentum crucis." By this move, Lyell altered the classic analogical argument for climatic change, criticized by Fleming, back to an inductive one. Put schematically, the change was from the analogical argument:

> Extant species A occurs in climate B at time T (the present) and location L.
> Extinct species A1 (in the same genus as A) occurred at time T1 and location L1
> _____
> Therefore, (probably) climate B1 (similar to B) occurred at time T1 and location L1.

to the inductive argument:

> Extant species A occurs in climate B at time T (the present) and location L.
> Extant species A occurred at time T1 and location L1.
> _____
> Therefore, (probably) Climate B occurred at time T1 and location L1.

Of course, the second argument rests on a suppressed premise that habitats of a given species do not change over time. In other parts of the *Principles*, Lyell tried to show that this was so. He relied on an argument, formulated by the Scottish philosopher Dugald Stewart: "The uniformity of animal instinct . . . pre-supposes a corresponding regularity in the physical laws of the universe" (Lyell

1830–33, 1:161). Consequently, quoting Stewart, "If the established order of the material world were to be essentially disturbed, (the instincts of the brutes remaining the same), all their various tribes would inevitably perish." *The uniformity of instinct was the foundation stone of Lyell's reconstruction of earth history.* Any naturalist, he said, will "admit that the same species have always retained the same instincts, and therefore that all the strata wherein any of *their* remains occur, must have been formed when the phenomena of inanimate matter were the same as they are in the actual condition of the earth." Using this principle, the geologist could explore the past because "the same conclusion must also be extended to the extinct animals with which the remains of these living species are associated; and by these means we are enabled to establish the permanence of the existing physical laws, throughout the whole period when the tertiary deposits were formed." Of course, critics such as Fleming could have replied that asserting the uniformity of instinct as a principle did not really solve the problem.

Lyell used the Tertiary strata of the Italian peninsula to reconstruct past climate. There, he found extant, but fossilized, species from different time periods within the Tertiary. Taking these in sequence, Lyell rendered more plausible the claim that temperature had declined during the fairly recent geological past. In the extreme south of Italy, Lyell had examined strata that had only recently been consolidated and uplifted from the bed of the sea to form land. He found that recently fossilized individuals of many species were almost identical with, or perhaps slightly larger than, those still found in the Mediterranean. He construed this to mean that the climate of the immediate past must have been very similar to that now prevailing, or perhaps slightly warmer, allowing the animals to reach their somewhat larger size. Further north in Italy, Lyell had found strata that contained successively larger proportions of extinct species and were therefore, he argued, progressively older. These extinct species allowed no "direct" proofs of the type of climate that had prevailed. However, the strata also contained a number of fossilized individuals of extant species, which did allow such inferences. Many of these individuals belonged to species now found predominantly in the warm waters of the Indian Ocean, suggesting that the Mediterranean climate had been warmer when these older strata were laid down. This conclusion was supported by the appearance of the individuals of species still found living in Mediterranean waters. Those fossilized in older strata were fine specimens, whereas those found living in the area were "dwarfish and degenerate," suggesting to Lyell that the climate of the Mediterranean, where they had formerly flourished, was now almost too cold for them to survive. For Lyell these facts amounted to a "demonstration . . . not neutralized by any facts of a conflicting nature" that climate had deteriorated during the Tertiary epoch (Lyell 1830–33, 1:94–95).

In central and northern Europe, where earth movement in the recent past had been rarer, geologists had found only freshwater lake and river deposits, dashing Lyell's hope of completing a sequence of marine forms stretching back into the past. Even so, as we have already seen, these freshwater deposits contained fossil mammals closely resembling present-day tropical mammals. The evidence from these extinct mammal species was consistent with the evidence from the extant mollusk species that Lyell had already discussed. It supported the view that the region's climate had once been warmer. In turn, this gave added weight to the argument from analogy, and Lyell (1830–33, 1:96) decided, "As far . . . as proofs from analogy can be depended upon, nothing can be more striking than the harmony of the testimony derived from the last-mentioned sources." During the Tertiary at least, Lyell concluded, the northern European climate had been warmer and its subsequent cooling had caused widespread species extinction.

The secondary rocks, which contained no extant species, still presented a problem. Indeed, Lyell, perhaps unknowingly following Blumenbach, defined the secondary strata as those that contained almost no extant species. Thus, he had only analogical arguments at his command when reconstructing the history of earlier periods. Lyell accepted this and produced no new arguments. He merely mentioned, rather briefly, that he agreed with other geologists that the flora and fauna of these strata resemble those now found in the tropics and that therefore the climate must have been equatorial. He made no attempt to extrapolate back in time beyond the oldest fossil-bearing rocks; to his mind there could be no empirical warrant for any speculation about the nature of climate prior to the appearance of fossils. Lyell (1830–33, 1:103) rested the matter by claiming that since that time, "from the considerations above enumerated, we must infer, that the remains both of the animal and vegetable kingdom preserved in strata of different ages, indicate that there has been a great diminution of temperature throughout the northern hemisphere, in the latitudes now occupied by Europe, Asia and America." Despite this weakening of the tail end of the argument, Lyell seemed to feel quite satisfied that he had decisively confuted

Fleming's position and given a solid methodological foundation for the use of fossils to reconstruct past environmental conditions as well as to identify different strata.

Geologists, like Lyell, were intensely concerned with the methodological foundations of their discipline. From the mid-eighteenth century on, they relied more and more heavily on fossils to reconstruct the history of the earth. At first, they assumed that this could be carried out strictly in terms of enumerative inductive inferences. But once they recognized that large numbers of species had become extinct, they realized that the inference, if it could be made at all, was analogical rather than inductive. This threatened the whole enterprise of historical geology. Some were prepared to accept the results of analogical inferences only in the case where a number of independent analogical inferences, using different genera, all pointed to the same results. Other geologists, such as Lyell, attempted to put the arguments back on an inductive footing, by finding species whose ancestry stretched far into the past. For Lyell, as for most other geologists, the evidence appeared compelling when considered in this light.

It is worth noting that geologists turned to contemporary biological laws or generalizations, not to *geological* ones, when using fossils to reconstruct past climates. In this way, the historical geologist, although requiring causal theory in order to reconstruct the past, did not require *geological* causal theory. The independence of strictly historical geology and casual geology that we referred to in the introduction was well on its way to being established. In chapter 8, we shall look at the independence of causal geology from historical geology.

THE WERNERIAN RADIATION: A PRE-LYELLIAN SYNTHESIS

In an important article, Rudwick (1971, 213) asked whether Lyellian geology had been "launched upon a pre-paradigm world?" He answered that it had not. Rather, it had been launched in opposition to a "directionalist synthesis" forged in the 1820s. The main elements of this synthesis, as Rudwick defined it (1971, 214–220), were a geophysical theory of central heat; a belief that the earth had cooled; a belief that the earth's living beings had shifted with the environment, particularly with changing temperature; a belief that life had shown increasing complexity and diversity over time; and, finally, a belief that the earth's history had been interrupted by sudden and dramatic events.

Although I would not choose the Kuhnian language, I am completely in agreement with Rudwick (and with his commentator, Rappaport) that geology was "scientific" before Lyell. I agree with Rappaport (1971) that geologists were committed to empiricism and the rule of law. However, I would like to suggest that both positions need modification in light of the scholarship of the last fifteen years. I agree with Rudwick, in opposition to Rappaport, that geologists had arrived at synthesis or theory before Lyell. But I want to reformulate Rudwick's characterization of that synthesis. With one important exception—the *geophysical* theory of the cooling earth—his characterization fits the theory or synthesis adopted by those geologists (and they were the majority of geologists) who were part of the "Wernerian radiation." All the planks of the directionalist synthesis that Rudwick describes were held by Wernerian geologists. In fact, the elements of this theory were in place well before the 1820s. The cooling earth, the relation of fossils and the environment, and the increasing diversity of animal life had been familiar themes since the late eighteenth century. Episodes of dramatic change appeared later in the second decade of the nineteenth century. The one plank of Rudwick's "directionalist synthesis" that Wernerians did not accept was a geophysical theory of central heat. As we shall see in the next chapter, Wernerians continued to base their causal theories on chemistry until the late 1820s.

To summarize, there was a generally held theory before Lyell, but its provenance is not mysterious, and one does not have to search through specialist monographs to piece it together. This theory did change over time, and geologists did subscribe to different variants of it; but then that, as I have argued, is the nature of science. The theory in question is the one that was taught in the major centers for geological training, the one that guided the research of most geologists—namely, the Wernerian theory as it developed in the generation of the Wernerian radiation.

8
Wernerian Causal Geology

The success of historical geology did not mean the demise of causal geology. Far from it. Wernerians, particularly Leopold von Buch, Alexander von Humboldt, and Karl von Hoff, all contributed substantially to the development of causal theory. The first two did so by their theories of elevation by heat. The third did so by his lengthy survey of the contemporary agents of geological change, written in response to a prize offer, made in 1818 at Blumenbach's suggestion, by the Royal Society of Sciences in Göttingen for the best "investigation of the changes that have taken place in the earth's surface confirmation since historic times and the application which can be made of such knowledge in investigating earth revolutions beyond the domain of history" (Hoff 1822). All three developed their causal theories largely independently of the strictly historical reconstructions of the earth's past in terms of formations and fossils that they were simultaneously undertaking. Then, at the end of the 1820s, the French geologist Léonce Élie de Beaumont put forward a genetic causal theory that finally displaced the older Wernerian one, incorporating some of the more important conclusions of historical geology and employing physical rather than chemical causes.

In the late eighteenth century, geologists had tried to solve three causal problems: the problem of the consolidation of rocks and minerals, the problems of the change of relative level of land and sea, and the problem of the erosion of the land surface (probably the least pressing of the three). Werner, in his genetic theory, had given a unified answer to all three problems in terms of an ocean from which rocks were deposited and which, by its intermittent retreat, exposed these rocks and occasionally eroded them. To the succeeding generation, his unified answer no longer seemed satisfactory. Nor did the traditional weighting of the three problems. Consolidation declined in importance. Erosion continued to attract only

limited interest. The changing position of the land and the sea moved
to the center of attention. Several kinds of evidence—the presence
of recent formations on mountaintops, folded and faulted strata, and
"denudations" had convinced geologists that something more than
variation in the level of the ocean was required. Most turned to
elevation by heat for an answer.

WERNERIAN THEORIES OF ELEVATION

Wernerians had never been averse to the idea that the primitive
ocean might have been hot. They had long known that many of the
rock-forming earths, particularly silica, were soluble only in hot,
basic liquids (see chapter 3). Chemists had shown that, contrary to
earlier theories, crystals could grow from melts as well as from
aqueous solutions. Furthermore, the fossils found in older rocks
suggested that the earth had once been hotter.

Two developments inclined Wernerians to extend their willingness to consider heat as a geological agent to wholehearted support
for the idea. First, field geologists argued that basalt, a widely
distributed and puzzling rock, had flowed from volcanoes, indicating
that igneous activity had been widespread in the past. Second,
chemists formulated theories of volcanic action that made it an
integral part of the history of the earth, rather than a minor,
epiphenomenon, as had been formerly thought.

Basalt is a dark, fine-grained rock that frequently occurs interbedded with stratified rocks in the geological column. We now assume
that it flowed from volcanic vents in the earth's crust, in a type of
vulcanism known as submarine or Hawaiian. Matters were not so
clear in the late eighteenth century; neither its place in the succession, nor its mineralogy, nor its stratigraphic relations, nor evidence
from contemporary volcanoes decisively indicated its origin.
Werner had been puzzled about basalt, but had treated the puzzle as
a minor one. Like many such puzzles in science, it loomed larger
and larger as time went on. Werner had originally placed basalt in
the primitive class, but later moved it to the *Flötz*, the stratified
rocks (Wagenbreth 1955; Ospovat 1971). Since it formed part of a
formation suite that also included porphyry-slate, wacke, and ironclay in the stratified class, and hornblends, greenstone slate, and
primitive greenstone in the primitive class, changing its place in the
succession did not indicate a major reevaluation of its nature
(Jameson 1808/1976; also see fig. 8).

Mineralogically, basalt appeared to be "indurated black-coloured ironclay," suggesting that it was an earth, probably an argillaceous earth (Jameson 1808/1976, 186). Earths were not the kind of mineral expected to flow from volcanoes because they were so resistant to heat. Basalt was stony in texture, again suggesting it could not have been formed by heat, but had probably been deposited by water (see chapter 3). Furthermore, basalt frequently formed large pyramidal columns—for example, at the Giant's Causeway in northern Ireland, and Fingal's Cave off the Scottish coast. Most mineralogists thought these were enormous crystals, again consistent with basalt's consolidation by water, since the columns could be considered either as crystals precipitated from water or as contraction figures formed by dessication (Taylor 1968). The French geologist Jean Étienne Guettard (1715–1786) echoed general mineralogical opinion when he described basalt as "a species of vitrifiable rock, formed by crystallization in an aqueous fluid," concluding that there was no "reason to regard it as due to igneous fusion" (Guettard 1770, 268).

The conviction that basalt's mineralogy indicated consolidation from water, which had seemed so self-evident in the 1770s and 1780s, weakened in the succeeding forty years. Mineralogists realized that stony and crystalline textures could be produced both by cooling from a melt and by deposition from water. Some geologists, such as James Keir, suggested that the basalt columns were evidence for an igneous, not an aqueous, origin (see chapter 3). Then, in the second decade of the nineteenth century, Louis Cordier (1777–1861) compared specimens from ancient lavas, among which he included basalt on the basis of field evidence, and equivalent lavas from modern volcanoes to see if he could detect any difference (Cordier 1816; Ellenberger 1984). He powdered the rocks, separated their components by flotation, and then analyzed the particles under the microscope, in the chemical laboratory, and with magnets. He discovered that the texture and mineralogy of both ancient and modern lavas were very similar, adding credence to the igneous origin of basalt. Even so, clear mineralogical evidence for the igneous origin of basalt did not become available for a couple of decades, until the technique of examining rocks in thin section was introduced. This finally showed the microcrystalline structure of the rock.

The stratigraphic evidence was even more confusing than the mineralogical. Older basalts usually occurred as conformable caps on top of hills of stratified rocks. The basalt over the chalk in Antrim in northern Ireland and the basalt on the Stolpen Mountain, which had decided Werner (1786/1971) in favor of an aqueous origin, were just two of the most famous examples.

But if some basalts appeared to be stratified, others appeared to have flowed from volcanic vents. In the Auvergne, basalt (some of it prismatic) occurred close to a series of curious cone-shaped mountains. Guettard, who had visited the area in 1751, had already argued that these mountains were extinct volcanoes (Guettard 1756; Geikie 1905/1962; de Beer 1962). But he did not conclude that the basalt had flowed from the volcanoes. That was left to Nicholas Desmarest (1725–1815), who visited the Auvergne in 1763 and 1766, studying the prismatic basalt of the area with some care (Taylor 1968). He traced what appeared to be flows of basalt back to the volcanic cones. On that basis, he claimed that basalts had a volcanic origin (Desmarest 1774). Where basalt appeared as an outcrop isolated from the rest of the flow, he argued that water had eroded the intervening basalt. But the recognition of the igneous origin of basalt did not imply that volcanoes were major causal agents. Indeed, he thought they were quite minor, stating that he had always believed "volcanic eruptions" to be "accidents among ordinary phenomena of nature" (Desmarest 1806, 221). So the stratigraphic indications of the origin of basalt were contradictory. To make matters yet more confusing, Werner and two of his pupils, Freiesleben and von Buch, the latter two in particular being known as excellent field geologists, reported having found fossils in basalt in a number of locations (Jameson 1808/1976, 187).[1]

Observations of active volcanoes did not decide the issue of basalt's origin or significance either. No European volcanoes spewed out a lava that resembled basalt. Active European volcanoes were few and far between, by far the most famous being Vesuvius and Etna on the Italian peninsula. Their major product, apart from ash and pumice, was a light-colored, viscous lava, then generically known as trachyte, which had quite different properties from basalt.

Most late eighteenth-century mineralogists thought that the balance of mineralogical and stratigraphic evidence tipped in favor the theory that basalt had been deposited as a stratified rock. In this, Werner followed Klaproth, Wallerius, and Bergman. He firmly defended the theory of the aqueous origin of basalt when Albrecht Höpfner carried a series of articles on the debate in his *Magazin für die Naturkunde Helvetiens* (1789).

But given the ambiguous nature of the evidence, many Wernerians opted for a volcanic origin for basalt. Blumenbach's friend Deluc had accepted basalt as volcanic by 1778. Johann Carl Wilhelm Voigt, who had been trained by Werner before becoming councillor of mines at Ilmenau in Thuringia, published a survey, complete with maps, between 1781 and 1785 in which he implicitly accepted that basalt had cooled from a melt. So did Raspe in 1763, Jean-Claude de Lamétherie (1743–1817) in 1797, and Karl von Hoff in 1810 (Carozzi 1969). But, with the exception of Raspe, these mineralogists continued to believe that most rocks were consolidated by water, and they attached little importance to volcanoes as a geological agent.

To see why this was so, we have to understand the chemical theories of volcanic activity. A few geologists believed that there was a central heat source in the earth that fueled volcanoes, making them a major causal agent. They included Robert Hooke, Lazzaro Moro, Buffon, Jean Bailly, Pierre-Simon Pallas, Jean Mairan, and Rudolf Raspe. But most geologists, particularly those in the Becher-Stahl tradition, disagreed. To them, the isolated location of volcanoes, their intermittent eruptions, their frequent association with recent rocks, and the differences between recently erupted trachyte and basalt spoke to a recent and superficial cause. So, too, did chemical theory. Most mineralogists postulated that volcanoes were caused by fires or fermentations just below the earth's surface (Taylor 1977). The fuel most commonly proposed was coal, followed by pyrites. Volcanic mountains were the site of volcanoes, but not created by them. And the products of volcanic fires were altered stratified rocks, not eruptions from the earth's hot interior.

For instance, Desmarest, who had argued that the Auvergne basalts flowed from volcanoes, nonetheless believed that volcanoes were recent and limited causal agencies. He attributed them to underground coal fires, and since the primitive rocks (or the *terre ancienne* of Desmarest's teacher Rouelle) contained no coal, volcanoes could have erupted only in recent times. Like others, he suggested that the heat created by these underground fires changed the water contained in rock pores into steam, and that the expansion caused explosive eruptions. Desmarest, like his contemporaries, did not connect the heat source with the ejected lava. Lava, he thought, was formed from stratified rocks that had been melted by the volcanic fires. Desmarest speculated that granite was the raw material that was altered to basalt, since those very properties that mitigated against granite having been formed by heat—its tendency to melt when heated and to cool to a noncrystalline slaggy mass— would be consistent with it forming basalt under the action of heat.

Werner (1789) claimed to have a new twist on the old theory that burning coal strata caused volcanoes—namely, a specification of the conditions under which fires would cause volcanoes. Three preconditions had to be satisfied: vast coal beds had to have been deposited; these had to have been covered by rocks that, when melted, would form lava; and water had to enter the burning coal bed to cause the eruption (presumably by the expansion of steam). Furthermore, the beds overlying the coal had to be basalt or "wacke"—usually a friable weathered basalt, as at the Meissner in Hesse and the Westerwald in Westphalia. He offered the hornblende crystals frequently formed in lava as proof, arguing that the groundmass of basalt was melted by the fire but not the hornblende, which therefore remained as crystals in the lava. The mineralogist Déodat de Dolomieu, who had perhaps observed more extant volcanoes than anyone else in the eighteenth century except William Hamilton, wrote that he was "certainly more in accord with [Werner] than with all the other French, Italian and English mineralogists. For, far from extending the empire of subterranean fire, I believe that more than any other person I have circumscribed its true limits and excluded many lands and substances from its domain" (Dolomieu 1790, 193; translated by Taylor 1968, 315).

Then, in the early nineteenth century, Humphry Davy (1778–1829) formulated a new theory of the origin of volcanoes that the Wernerians found very appealing. Among those who welcomed it were Robert Bakewell (1813), Charles Daubeny (1826), and Alexandre Brongniart (1816–30). The theory which was squarely in the chemical and mineralogical tradition, called neither for an internal fire nor for a rigid crust floating on a fluid interior, and it could explain why volcanic activity appeared to have been greater in the past.

In 1806, Davy had announced that he believed that chemical attraction was electrical in nature (Siegfried and Dott 1980). Using electricity, the chemist ought to be able to break apart compound bodies into their true elements. By 1807, he had isolated both potassium and sodium. The following year he added calcium, magnesium, barium, and strontium. Many of the earths, then, were not simple, elementary bodies, as chemists and mineralogists had assumed when Werner was first writing and when Lavoisier was proposing his reforms of chemistry. Instead, they were compounds of a metal and an air, oxygen. When combined with water, all these metals produced heat and removed the oxygen from the water, producing (in modern terms) the hydroxide of the metal and hydrogen.

Davy had shown that the earths, limestone and dolomite, far from being elementary and inert, were combinations of metals with water or acids. Mineralogists had long believed that the interior of the earth was rich in metals. If these included native calcium and magnesium (and possibly as yet undiscovered metals) and if water penetrated the interior of the earth from time to time, violent chemical reactions and hence volcanic explosions could be expected to follow.

Indeed, Davy was in one sense just reviving theories that had been around for some time (Taylor 1977). Several late-eighteenth-century scientists, including Lavoisier and Kielmayr, had toyed with the idea that the alkaline earths might be metallic oxides (Partington, 4: 45). And von Hoff (1822) reported that a few years previously, Werner's pupil Heinrich Steffens had made the oxidation of the hypothesized metals the basis of his theory of terrestrial heat.

Although Davy never published a detailed formulation of his theory, he described it at length in his popular public lectures. "The interior of the globe," he said, was "composed of the metals of the earths, which the agency of air and water might cause to burn into rocks; and even the reproduction of these metals may be conceived to depend upon electrical polarities in the earth" (quoted in Siegfried and Dott 1980, xxxix). The heat given off when the metals combined with air and water fueled the igneous geological processes taking place in the globe. The old puzzle of the source of the oxygen (or air) needed for combustion within the earth had been solved by the earlier discovery that water was composed of hydrogen and oxygen. Hence, the rivers and oceans of the world were now seen as a bountiful source of oxygen. Indeed, Werner's emphasis on the centrality of water as a geological agent was confirmed by this development.

Davy even produced an experimental demonstration of volcanic action, perhaps mindful of the oft-cited model of Nicholas Lemery (1645–1715) of a volcanic explosion using sulfur and pyrites. An observer reported,

> A mountain had been modelled in clay, and a quantity of the metallic bases introduced into its interior; on water being poured upon it, the metals were soon thrown into violent action—successive explosions followed—red hot lava was seen flowing down its sides, from a crater in miniature—mimic lightenings played around, and in the instant of dramatic illustration, the tumultuous applause and continued cheering of the audience might almost have been regarded as the shouts of alarmed fugitives of Herculaneum or Pompei. [Siegfried and Dott 1980, xxxix–xl]

In 1813, Davy set out on a two-year tour of the Continent, largely to test in the field his new hypothesis about the origin of volcanoes.

Joseph-Louis Gay-Lussac (1778–1850), who had visited Vesuvius with Humboldt and von Buch in 1805–6, and who was an acquaintance and rival of Davy's, suggested some modifications of Davy's theory. He published his own widely reprinted version in 1823. According to Gay-Lussac, the gases emitted by volcanoes were not, as Davy proposed, pure hydrogen, but chlorides and sulfides. Consequently, Gay-Lussac suggested that the water feeding the volcanoes must be salt, not fresh, and the metallic interior of the globe must contain some of their metallic compounds.

In 1818, shortly after Werner's death, Leopold von Buch (1774–1853) publicly announced at the Berlin Academy a theory of volcanic action with which he had been playing for some time—namely, the theory of "elevation craters" (Buch 1820). He argued that basalt was responsible for the uplift of substantial areas of land, using Davy's theory to account for its action. He followed up this short paper with an impressive monograph on the Canary Islands (Buch 1825). In it he laid out his evidence for elevation craters at even greater length.

Von Buch was as Wernerian as anyone (Dechen 1853; Wagenbreth 1953a, 1953b; Mathe 1974; Guntau 1974b, 1974c). After three years studying with Werner and a further year studying law and government at Halle and Göttingen, he joined the Prussian civil service as a mine inspector in 1796. Shortly after, he took advantage of an inheritance to finance his own geological travels for the rest of his life, though not before surveying Silesia.

Von Buch had visited Vesuvius as a young man in 1798. In spite of a determined search, he had failed to find either coal or pyrites that could have fueled the volcano. In 1802, a visit to the Auvergne convinced him that the "trachyte" mountains there had once flowed as lava. The trachyte lava, he thought, was altered granite (Buch 1867, 1:486–87). This was quite consistent with the general Wernerian theory that lavas were altered products of rocks, such as granite, that had been consolidated by water. The Auvergne rocks were sufficiently similar to those found on Vesuvius to convince von Buch that they originated in a volcano; they were sufficiently dissimilar to raise a series of problems. Von Buch wondered how distinct, narrow lava flows could form a symmetrical cone like Vesuvius. He wondered how smooth craterless domes, like the Puy de Dôme in the Auvergne, could have been caused by the eruption of lava, ashes, and pumice through a vent. He wondered whether volcanoes could be the final stage of igneous activity, following the elevation of superincumbent strata to form a cone.

In 1805–6, von Buch visited Vesuvius a second time, accompanied by Gay-Lussac and Humboldt. The trio had the good fortune to observe a full eruption. We may assume that Humboldt told his companions how this compared with the volcanoes he had recently observed in Mexico. On his journey through South and Central America in the years 1799–1804, Humboldt had visited the impressive Mexican volcano Jorullo (Baumgärtel 1959). This had erupted as recently in 1759. It was "a mountain of scoriae and ashes, 517 met. [563 yards] above the old level of the neighbouring plains, suddenly formed in the centre of a thousand small burning cones, thirty-six leagues from the sea shore, and forty-two leagues from any other volcano" (Humboldt 1810, 81). The ground remained warm, and eyewitnesses lived to tell the tale.

Humboldt was persuaded that the plain surrounding the volcano's focus had swollen, like a giant bubble blown up from below, to form the gently conical shape known as the "malpais." Humboldt (1810, 86) was convinced that "geological reasoning can be supported only on the analogy of facts that are recent, and consequently well authenticated." Consequently, he was delighted to find a case of recent volcanic action. It suggested that local formations had been elevated by igneous forces acting from below. Humboldt and von Buch always cited Jorullo when challenged to produce contemporary evidence of major land elevation by igneous causes, since they believed Jorullo constituted "one of the most extensive physical changes, that the history of our globe exhibits" (Humboldt 1810, 81).

Nor was this unusual. Humboldt believed that the Jorullo volcanic eruption was causally connected with simultaneous events in the Cordilleras and the Andes. It showed how much more extensive volcanoes were than their surface effects indicated and served to dispose of Cuvier's conclusion that volcanoes were limited and insulated phenomena. Humboldt thought it likely that volcanoes had caused widespread elevation because all the evidence suggested that they were interconnected.

After visiting Vesuvius, von Buch went off between 1806 and 1808 to observe the famous phenomenon of the changing level of land and sea around the Baltic. Although the theory that the changing level resulted from a shrinking ocean had been commonplace amongst eighteenth-century Swedish geologists, von Buch argued that the

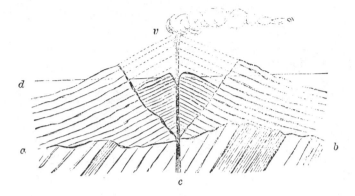

Fig. 16. Hypothetical section through a crater of eruption (De la Beche 1851, 319). The walls of the crater and the interior cone have been formed by the products of volcanic eruptions.

land was rising, not the sea shrinking.[2] By 1813, von Buch had carefully studied the mineralogy of many of the rocks most closely associated with vulcanism. One was the light-colored lava, trachyte, found in many contemporary volcanoes and in the extinct volcanoes of the Auvergne. The other was porphyry, a rock with large crystals in a fine groundmass (Buch 1816). He began to distinguish between true volcanoes and wider domelike crustal elevations of the type found in parts of Hungary and the South America Cordilleras.

In 1815, von Buch visited the Canary Islands. On his return, he was ready to deliver his paper on elevation craters (Buch 1820). He distinguished two forms of igneous activity. Each had a distinct type of cone and crater as well as other structural, morphological, and mineralogical features. True volcanoes (or craters of eruption) were exemplified by Vesuvius; basalt islands (or craters of elevation) were exemplified by the peak of Tenerife.

True volcanoes, on von Buch's analysis, were isolated, relatively small cones (fig. 16). They were composed largely of viscous, rapidly solidifying, light-colored trachytic lavas, dominated by feldspars. Volcanic cones were built up by the solidifed lavas emitted through the central vent. In addition, smoke, vapor, and solid matter, such as ashes and pumice, were emitted and the latter fell as unsorted deposits on the sides of the cone.

Basalt islands, or craters of elevation, were more typical and more important forms of igneous activity (figure 17). They were much larger, and generally found in or near the ocean. The distinctively

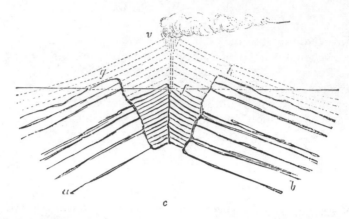

Fig. 17. Hypothetical section through a crater of elevation (De la Beche 1851, 319). Volcanic forces have broken through the superincumbent rocks at (c) (rocks that in this case happen to be stratiform, through they might also be crystalline), elevating them on both sides. A secondary crater of eruption (f) has been formed of volcanic products inside the crater of eruption.

igneous component, dark-colored basalt, was usually interbedded with ordinary sedimentary strata. These islands rarely contained the trachyte lava flows and unsorted ejected materials characteristic of true volcanoes. As von Buch knew, compared with trachyte, basalt flows freely and solidifies slowly. Therefore, he rejected the idea that it had solidified on the slopes of a volcano. He argued that it had flowed from a vent on the ocean floor, spreading over a wide area and solidifying deep under the sea. Sediment collected on top of the basalt flow and solidified, only to be covered later by another basalt flow, until a thick layer of interbedded igneous and sedimentary rocks had been built up on the ocean bed. This explained the curious stratigraphic relations of basalt.

Von Buch speculated that, if the vent through which the lava flowed was blocked, then the whole surrounding area would be forced up by the steam and gases, or possibly by the intrusion of molten material below the surface. The layers of basalt and sedimentary rock would form an island with strata sloping gently away from the central pressure point. When the elevating force was strong enough, the strata would split apart to form a central kettle-shaped crater. The size of the elevation craters, their depth, the absence of ordinary lava flows, and the presence of marine strata all combined to convince von Buch that these islands must have been elevated by some force deep in the interior of the earth.

True volcanoes could sometimes be found in the center of elevation craters. Elevation craters, after all, were weak spots through which molten material and volcanic products could easily escape. When this happened, the expansive force would be released and no further elevation would take place. The volcano Vesuvius, for example, was located within the elevation crater Somma.

The vulcanism was caused in the way Davy and Gay-Lussac had suggested. Von Buch argued that the oxidation of metals by the accession of sea water caused elevation craters (Buch 1867, 2:64–68). But although the chemical theory of vulcanism nicely explained von Buch's theory of elevation craters, the elevation crater theory could survive without its chemical underpinnings. Von Buch had identified a distinction in the structure and mineralogy of different kinds of volcanoes that demanded explanation by anyone who sought to discredit him.

Elevation craters could be either "central" and "serial." Central elevation craters occurred at local points of crustal weakness. For this reason, Bernard Studer (1794–1887) named the major peaks of the Alps "*massifs centrals*," a name that has lingered on long since the theory that justified it vanished (Studer 1834). Serial elevation craters, usually along a line of weakness, accounted for *regional* uplift.

Von Buch and Humboldt believed that the intensity and mineralogy of volcanoes changed over time. "Volcanic matter, which during thousands of years has been progressively raised towards the surface of our planet in such different circumstances of mixture, pressure, and cooling, must display both contrasts and analogies. . . . The nearer we approach to modern times, the more the volcanic formations appear insulated, superadded, and foreign to the soil over which they are spread" (Humboldt 1823, 411–12).

For the next thirty or forty years, the elevation crater hypothesis was the dominant theory of vulcanism. "A modified volcanic theory" had won the day, firmly announced the president of the Geological Society of London, citing the work of Hutton, von Buch, and Humboldt (Fitton 1827-28, 55-56). Some of theory's supporters were figures with whom we are already familiar; others were younger geologists. Humboldt, Omalius d'Halloy, Hoff, Élie de Beaumont, Johann Steininger (1794–1874), who worked on the volcanoes of the Rhine, Wilhelm Abich, who studied Etna and Vesuvius, Landgrebe and Henri Sainte-Claire Deville (1818–1881), and (in the early stages) Constant Prévost and Friedrich Hoffmann all subscribed to its main doctrines. Charles Daubeny (1826) described the elevation crater theory at length in his textbook on volcanoes. William Buckland used the theory to explain the "den-udations" (eroded anticlines) in southern England (Buckland 1829).

The theory of elevation craters received additional support from strange events that occurred in the Mediterranean in 1831. After some preliminary signs of volcanic activity in late June, a volcano began to appear above the water in mid-July, reaching a maximum size of about 3,000 feet in diameter and 100 feet in height, before vanishing within the year (Dean 1980). It was given a variety of names by a variety of nationalities, the commonest being Graham Island and Île Julia. Most geologists saw this as the creation of a crater of elevation.

However, in 1835, two geologists who had seen the volcano, Friedrich Hoffmann and Constant Prévost (1835), expressed their doubts about whether it really was an elevation crater. In the same year, Lyell and Prévost met von Buch and Élie de Beaumont (who had recently revisited the Italian volcanoes) at the meeting of the German Association for the Advancement of Science in Bonn. Lyell reported to Mantell on 14 October 1835 (Lyell 1881, 1:456),

> For two whole days Constant Prevost and I fought Von Buch and
> de Beaumont on the craters of elevation, the discussion being in
> French, and with a crowded audience. I am as convinced as ever
> that their views are quite erroneous, and if I had to write over again
> to-morrow my chapter on that subject, I should have nothing to
> retract and many new arguments to add.

Lyell and his friend George Poulett Scrope (1797–1876) had long had doubts about the theory of elevation craters. Scrope, who had studied the volcanoes of the Auvergne, allowed that Humboldt and von Buch had correctly noted that trachyte usually occurred as masses rather than as sheets. But he scornfully dismissed them for having asserted that elevation craters "swelled up like a bladder by inflation from below, and are consequently still *hollow within*—a gratuitous supposition entirely at variance with all that we know for certain concerning the nature and mode of operation of the volcanic energy" (Scrope 1825, 91). Lyell had probably heard about the theory in Buckland's classes in Oxford. He would certainly have encounted it when he met Humboldt in Paris in 1823, when Humboldt recommended Hoff's compilation (1822) as the best documentary account of earthquakes and volcanoes. Lyell charged that von Buch had underestimated contemporary causes and that his theory did not stand up to the test of observation. There was no known "crater formed exclusively of marine or lacustrine strata, without a fragment of any igneous rock intermixed," as might be expected if elevation craters were formed without the primary agency of eruptions (Lyell 1830–33, 1:387).

Von Buch was unswayed. In 1835, at the Royal Academy in Berlin, he charged that Lyell's observations were unreliable and reasserted his unwavering commitment to mountain building by elevation craters (Buch 1836). And indeed, Lyell himself had reservations that never made their way into the *Principles*, for he thought it possible that there might be some upheaval and disruption of the strata during volcanic eruptions (Dean 1980).

In the 1840s, the elevation crater hypothesis was still drawing expressions of support. Humboldt reasserted his belief in elevation craters in *Kosmos* (1845–1862), which was translated into English in 1845–58. Charles Darwin came out strongly in favor of them in one of his more popular works, the *Geological Observations on Volcanic Islands* (1844/1897). For several years after returning from the Beagle voyage, Darwin worked on the problem of elevation (Darwin 1837a, 1837b, 1840, 1844/1897; Herbert 1970, 1980; Rudwick 1974a). He agreed with von Buch that craters of elevation were distinct from ordinary volcanoes. St. Helena, St. Jago, and Mauritius, all largely composed of basaltic lava flows, "come into the class of craters of elevation," he declared (Darwin 1844/1897, 106). Darwin thought that Élie de Beaumont's observations had demonstrated, *contra* Lyell, that lavas could not consolidate on steep slopes.

The debate by now turned on whether all lava flows could consolidate on steep slopes, as Lyell insisted, or whether certain kinds of lavas tended to spread out before consolidating, and must have been elevated since, as von Buch believed (Lyell 1850). After von Buch's death in 1852, Lyell had one last try at convincing the scientific world that he had incontrovertible evidence that lava solidified on steep slopes (Lyell 1858). A translation was quickly authorized by the Geological Society of Berlin. Paris proved more difficult. Lyell approached the Swiss scientist Auguste de la Rive (1801–1873). He said that a friend in Paris had

> strongly recommended me to try to get it [the paper] done at Geneva—he remarks that the great difficulty in Paris is the influence of Elie de Beaumont, whose facts as well as his theory of Etna are attacked, and that on a question on which he is deeply committed. He has been as you will see throughout my paper the great champion of von Buch's hypothesis. [Lyell to de la Rive, University Library, Geneva]

European geologists held fast to their own traditions.

In the meantime, von Buch had put forward a supplementary theory of elevation, his theory of dolomitization. Although it did not achieve the success of the theory of elevation craters, it attracted a sizable following for a number of years. The theory of elevation craters offered a plausible explanation for the origin of *local* relief features such as Tenerife or Mount Somma. Since basalt flows were known to be widespread in many parts of the world, they might explain the elevation of areas as large as, say, Northern Ireland. Arranged in chains, they might even account for mountain ranges. But von Buch felt, perhaps because major mountain chains contained no basalt, that some other explanation for mountain formation was required. When Brongniart (1821) and Buckland (1821) used fossils to prove Ami Boué's theory that many of the rocks of the Alps were not primitive, as Werner had assumed, but of recent origin, the problem became particularly pressing.

By 1822, von Buch announced that he had found the "key" to mountain elevation in the Fassa Valley of the southeastern Alps. Geologists already believed that the region was significant. Thirty years previously, in 1791, Dolomieu had shown that the white crystalline peaks that towered around the valley were not marble, as appeared at first sight (Carozzi and Zenger 1981). Samples of this crystalline rock did not effervesce when tested with a weak acid, as marble (calcareous earth) should. By the 1790s, it will be recalled, the major earths were distinguished by chemical tests, and the difference between magnesia and calcareous earth was appreciated.

Dolomieu's announcement caused a stir in Paris, where the influence of Romé de l'Isle and later of Haüy meant that most chemists and mineralogists held that *crystal form* was the chief guide to mineral identification. Yet this rock, which was composed of small crystals that belonged to the same "hexagonal-rhombohedral" class as calcite, had chemical properties quite different from calcite. The following year the younger Saussure named this portion of the Alps the "Dolomites," and the rock in question "dolomite."

Intrigued by Dolomieu's work, the English chemist Smithson Tennant (1799) reported to the Royal Society that lime applied as fertilizer by farmers had two very different effects: sometimes it impeded or even killed plant growth; sometimes it stimulated it. Tennant analyzed samples that had caused each effect. He concluded that there were two types of limestone; one (that impeded growth) contained magnesia, and the other (that stimulated growth) contained calcium. Tennant showed that the former had been made from a rock similar to "dolomite." The major English deposit of this limestone soon became known as the "magnesian limestone." In 1821, William Buckland summarized what was known about the English formation, compared it with Brocchi's description of the formation in the Fassa Valley, and argued that the "alpine limestone" was contemporaneous with the English magnesian limestone.

Von Buch (1825) incorporated this mineralogical discovery into a theory of the genesis of mountain chains. He assumed that the intrusion of igneous rocks with accompanying steam and vapors caused elevation. In the Fassa Valley, the obvious candidate for the role of intruded rock was the large mass of "porphyry"—a fine-grained igneous rock containing large individual crystals. Von Buch proposed that just as basalt upheaved isolated elevation craters, so porphyry upheaved mountain chains, splitting open valleys and shattering rock that accumulated as conglomerates.

Von Buch distinguished "red porphyry" from "black porphyry." The former rock was closely related to granite and composed predominantly of quartz. He suggested that it had been formed prior to the black porphyry, which cut through it at intervals. Red porphyry, he thought, had little to do with mountain building. Black porphyry, on the other hand, he held to be the major cause of uplift. Black porphyry corresponded to the "trap" rock of the English geologists. Von Buch suggested that it be called "augite porphyry" after its chief distinguishing ingredient, the mineral augite.

Chemical analysis had shown that "magnesia" was an important constituent of augite. Dolomite also contained "magnesia." This coincidence was the key to von Buch's puzzle. He argued that the dolomite had originally been deposited as calcareous limestone. Subsequently, the magnesia associated with the augite porphyry had transformed it into dolomite. Some field evidence supported this. Dolomite regularly contained patches of pure calcareous limestone.

Chemical theory, as we might expect, played a role, too. Just a couple of years earlier, von Buch had heard Mitscherlich, a fellow member of the Berlin Academy, read his classic paper on isomorphism to the assembled company. Among other examples of isomorphous minerals, Mitscherlich had included the carbonates of lime (calcspar, or limestone) and magnesia (Mitscherlich 1818–19; Melhado 1980). Calcareous limestone exposed to some form of magnesia could presumably be transformed into dolomite by isomorphous substitution. Von Buch argued that the originally calcareous alpine limestone had been "dolomitized" by magnesia vapors released when augite porphyry was interjected. The vapors penetrated the

limestone through rifts created by the intrusion of the augite porphyry. In the course of dolomitization, the Alps had been upheaved.

Von Buch quickly generalized the theory. He visited the mountainous regions of the Thüringer Wald and the Harz, finding both
dolomite and augite porphyry, and suggesting that these mountains,
too, had been upheaved during the dolomitization of the local
limestone. He asserted that the intrusion of augite porphyry was the
cause of *all* mountain uplift. In some regions, the augite porphyry
was not exposed at the surface, so geologists had not recognized its
role in elevation as quickly as they might have done. Von Buch
believed that some other puzzling rocks might be the result of the
action of igneous vapors on limestone. With Charpentier, for
example, he suggested that gypsum might be the result of the action
of sulfur vapors on limestone.

Many geologists found the theory of mountain uplift by dolomitization intriguing. Henry De la Beche (1796–1855), for example,
claimed to have confirmed the theory by his observations on the
succession near Nice in the south of France (De la Beche 1834).
Adam Sedgwick (1785–1873), Woodwardian professor of geology at
Cambridge, explored the connection between the magnesian limestone in the north of England and the trap rocks of the area, even
though he expressed doubts about the details of the theory
(Sedgwick 1827, 1831a, 1831b., 1831d, 1835). Lyell was sufficiently
curious about the theory that in 1832 he proposed a visit to the Fassa
Valley to test it, although he eventually ended up going to the Eiffel
District instead (Lyell 1881, 1:396). By contrast, Daubeny (1826,
420) the leading British expert on volcanoes, and sympathetic to the
elevation crater hypothesis, roundly condemned upheaval by
dolomitization as "gratuitous in its assumptions, not very compre-
hensible in its details, totally inapplicable to most of the cases it is
intended to explain, and unnecessary to be resorted to in any."

By the mid- to late 1830s, most geologists agreed with Daubeny.
Two kinds of objections were advanced against the theory, one
chemical and one geological (Hoffmann 1838). The chemists of the
day, led by the prestigious Swedish chemist Jacob Berzelius
(1779–1848), argued that magnesia could not sublimate; therefore,
the suggestion that magnesia "vapors" caused dolomitization was
absurd. Geologists worried about the absence of augite porphyry in
some periods or places associated with mountain building, the lack
of alteration of limestone in some areas where augite porphyry was
found, and about the possibility that magnesian limestone was an
ordinary sedimentary formation.

Before leaving von Buch's causal theories, it is worth noting a few of their features. They depended on observations of current processes, on speculations about the earth's interior, and on chemical theory. They did not depend on the growing body of knowledge about the history of the earth, except the part that suggested that basalt flows had been commoner in the past and that mountains had been upheaved. The entities that von Buch used to construct his theories were the traditional ones of rocks and minerals, although the increase in chemical knowledge meant that they were described somewhat differently. Formations figured not at all. Causal theory was therefore independent from historical geology. But toward the end of the 1820s, a different kind of elevation theory was put forward; we must now turn to that.

ÉLIE DE BEAUMONT'S SYNTHESIS

In 1829, Élie de Beaumont announced a new account of elevation, employing physical, not chemical theory (Lawrence 1977; Greene 1983). Highly successful in the École Polytechnique, Élie de Beaumont had taken the unusual step of entering the Corps de Mines. During the 1820s, he earned his geological spurs with detailed surveys of the mines and mountains of the world.

Élie de Beaumont (1831, 241) thought that "the independence of the secondary formations is the most important result obtained from the study of the superficial beds of our globe," thus placing himself firmly within the tradition of Wernerian historical geology. Unlike some of his contemporaries, he did not doubt that there were sharp divisions between formations. His theory of elevation had to accommodate two well-established conclusions. The first, which he attributed to Cuvier, was that there had been "a succession of violent revolutions" in the earth's past. The second, which he attributed to von Buch, was that mountain chains had been elevated "by forces acting from beneath." Since we have just examined von Buch's theory of elevation craters, all we need to add here is that this was subsumed by Élie de Beaumont as a detail of his more global theory.

Cuvier had been the most prominent proponent of the thesis that the past was marked by violent revolutions. The frozen elephants and rhinos found in Siberia seemed to require a cause "as sudden as its effect" (Cuvier 1817/1978). So did the "breaking to pieces and overturning of the strata" and the "heaps of debris and rounded pebbles which are found in various places among the solid strata." Cuvier surveyed the "four causes in full activity, which contribute to make alterations on the surface of our earth" and came to his famous conclusion that it was no longer possible to "explain the more ancient revolutions of the globe by means of these still existing causes." He confidently concluded, "The thread of operation is here broken, the march of nature is changed, and none of the agents that she now employs were sufficient for her ancient works" (Cuvier 1817/1978, 24).

To account for the phenomena noted by von Buch and Cuvier, Élie de Beaumont distinguished twelve different systems of European mountain chains. He described their geography in detail. He was following well-established tradition. Werner, in his *New Theory of Veins* (1791), had proposed that mineral veins deposited at different epochs had different orientations. Humboldt had a yet more ambitious theory. He speculated that the rocks constituting the mountain chains formed in a given epoch would all have the same strike, and that the strike differed from epoch to epoch. At first he suspected that this might be connected with the earth's magnetism. On his trip to South and Central America, undertaken in part with the idea of testing this hypothesis, Humboldt was disappointed. He retreated to trying to find some geographical pattern in the mountains themselves, rather than in the strike of the rocks of which they were composed. Von Buch himself surveyed the German mountain systems, tracing their different orientations (Buch 1824).

Élie de Beaumont's innovation was to determine the period when each of these chains had been upheaved by noting the difference in age between the upheaved formations and the formations on the mountain flanks. This is a procedure quite different from the one von Buch had used in contructing his theory of elevation. We are back in the world of historical geology, where what matters are formations and ages, not mineralogy and chemical causes.

Élie de Beaumont concluded that the uplift had been sudden by geological standards. He argued that there had been twelve episodes of elevation, with accompanying revolutions. As mountains were suddenly elevated above the sea, floods would wash down their sides, and whole species and genera would be wiped out. Thus, the old legends of floods and deluges probably had a basis in fact. Cuvier was correct in thinking that there had been revolutions in the past. Furthermore, he had an answer to the long-standing problem of the "independence of formations." In conclusion, Élie de Beaumont (1831, 262) stated that he had achieved "one of the principal objects of my researches"—namely, he had shown "that this great fact [the independence of formations] is a consequence, and even a proof, of the independence of mountain-systems having different directions."

But this "positive geology" was not enough. "The mind would not rest satisfied," said Élie de Beaumont (1831, 263), "if it did not perceive, among those causes now in action, an element fitted from time to time to produce disturbances different from the ordinary march of the phaenomena which we now witness." Volcanic action seemed the obvious answer, but only if it was broadly defined in terms of the "influence exercised by the interior of a planet on its exterior covering during its different stages of refrigeration."

This was an idea that had become newly popular in the previous few years as the result of a series of developments in astronomy, climatology, and physics that went back some time further. The old idea that the earth had started as a molten body and slowly cooled had been given new life in the late eighteenth and early nineteenth centuries by a combination of developments in astronomy and physics (Lawrence 1977). Laplace worked out a mathematical and theoretical framework for the origin of the solar system, which he published as his great nebular hypothesis (Laplace 1824; Numbers 1977). Simultaneously, but independently, William Herschel provided observational evidence for this hypothesis (Herschel 1811). Noticing that different stars were more or less "nebulous" and of different colors, he postulated that they represented different stages in the evolution of stars from an initial, hot, nebular, rotating mass through denser and colder stars to old "dead" stars. But although the nebular theory created great interest in physics and astronomy, it seemed of scant relevance to geology, which, as we have seen, was busy with its own substantial agenda.

The issue came closer to home when the French physicist Joseph Fourier (1768–1830) suggested an explanation for the well-documented phenomenon that the temperature in mines increased with depth. The correct way to interpret this, Fourier argued, was as a "geothermal gradient" produced by the earth's remaining reservoir of original, internal heat (Fourier 1819). Louis Cordier (1827) took the geological side of the argument further. He calculated that the earth was fluid from about 5,000 meters below its surface, and argued that it was a cooled star that had consolidated from the outside in.

This was consistent with what was known about past and present climates. In the eighteenth century, long before any *historical* studies of climate has been carried out, scientists had realized that the temperatures recorded at present on the surface of the earth could not be explained solely in terms of incoming solar radiation. They were simply too high (Humboldt 1820). Numerous solutions were advanced to explain this anomaly. One of the favorites was that the central heat of the earth kept the surface warmer than would be expected if incoming solar radiation were the only heat source. Jean Mairan (1678–1771) in particular argued that "emanations from central heat" were essential to balance the earth's heat budget. In the early nineteenth century, these climatological considerations were reinforced by the fossil evidence that the earth's temperature had once been higher (see chapter 7). These various lines of evidence all pointed to the possibility that volcanoes were passageways to the molten interior of the earth and that many rocks might be the solidified remains of this interior. Élie de Beaumont proposed that, as the earth cooled and contracted, stresses built up in the crust. The sudden release of these stresses at infrequent intervals caused the periods of revolutionary mountain elevation as the crust buckled.

As the physical account of the source of volcanic heat grew in popularity, the chemical theory declined. By the end of the 1820s, Davy had severe doubts about his chemical theory of volcanic action. Although there appeared to him that "no other adequate source than the oxidation of the metals which form the bases of the earths and alkalies, . . . it must not be denied that considerations derived from thermometrical experiments on the temperature of mines and on sources of hot water, render it probable that the interior of the globe possesses a very high temperature; and the hypothesis of the nucleus of the globe being composed of fluid matter, offers a still more simple solution of the phenomena of volcanic fires than that which has just been developed" (Davy 1828, 250).

Élie de Beaumont's theory is much closer to the old tradition of genetic causal theories than to the causal theories put forward by his immediate, Wernerian predecessors. He did not determine his causes by examining present-day processes; indeed, he could not have done so, for we do not live in a period when mountains are upheaved in an instant. Rather, he derived them from physical theory. His theory of elevation is a hypothetical likely story to account for the new data accumulated by historical geologists. In this way, Élie de Beaumont synthesized historical and causal geology once again, at least for a time.

With the advent of Élie de Beaumont's new synthesis for geology, the old Wernerian program was well and truly dead. The synthesis was greeted with enthusiasm by geologists (Greene 1983). The program that had directed geological research for half a century was replaced by one with different foundations, a different attitude toward causes, and a much more catastrophist interpretation of the history of the earth.

9
Lyell's Geological Logic

Between 1830 and 1833 Lyell published the three large volumes that form his *Principles of Geology: An Attempt to Explain the Former Changes of the Earth's Surface, by Reference to Causes Now in Operation.* For many years, the work was regarded as the foundation stone of modern geology (see, e.g., Bailey 1962; Wilson 1972). More recently, there has been something of a backlash against this interpretation, several scholars arguing that Lyell's centrality to the development of geology has been much overplayed (Lawrence 1978; Greene 1983). That Lyell's book was widely discussed there can be no doubt. That his particular brand of geology shaped later developments is much more questionable. I shall argue that Lyell was chiefly concerned with the appropriate methodology for *causal* geology, making his work tangential to the overwhelmingly historical emphasis of his British contemporaries. And his choice of the appropriate method for causal geology ruled out most of the causal theories being advocated on the Continent, making his advice unacceptable to his foreign contemporaries.

Perhaps the best place to begin is with Lyell's choice of title. For scientists in the eighteenth and nineteenth centuries the word *principles* carried a heavy historical burden. Just as Newton had called his magnum opus the *Principia (The Mathematical Principles of Natural Philosophy)* in order to challenge the preeminence of Descartes's *Principles of Philosophy*, so later writers who chose to title their work *Principles* did so in the hope of emulating Newton's achievement. Lyell was no exception, and his ambitions were large. But what did emulating Newton's achievement mean to Lyell? One thing that it might have meant was linking geology firmly to the corpus of Newtonian theory, or at least to one of the many versions of it current in Lyell's day. That was the strategy adopted by many of Lyell's predecessors and contemporaries. Buffon, Bergman, Hutton, and Kirwan, among others, had already attempted to apply

their interpretation of Newtonian matter theory to geological entities. Lyell proceeded quite differently. He did *not* try to apply one
version or another of Newton's substantive theories to geology.
Instead, he argued that geologists should adopt the scientific *methods* advocated by Newton, for only in this way could geology
achieve the status of sciences like astronomy and mechanics. Lyell
intended to "establish the principles of reasoning" in geology by
becoming a "geological logician" (Lyell to Murchison, Lyell 1881,
1:234).

LYELL AND HUTTON

To understand why Lyell, more than any of his contemporaries,
believed that epistemic reform was necessary in geology, we need to
understand his perception of the state of geology in the 1820s
(Wilson 1969, 1972). As a young Scot studying law at Oxford,
Lyell's enchantment with geology waxed as his enthusiasm for his
official curriculum waned. His was one of a number of students who
found the informal lectures on geology by William Buckland a
breath of fresh air in the stuffy atmosphere of Oxford. Lyell spent
more and more of his time on geology, eventually abandoning law on
the grounds that his eyesight was not up to the strain.

   During the period between leaving Oxford in 1819 and drafting the
Principles in the late 1820s, Lyell became an advocate of the
Huttonian system. Textual evidence suggests that he encountered it
in Playfair's *Collected Works*, published in 1822 (K. J. Tinkler,
personal communication, September 1984). At first sight, the
Huttonian system was an improbable choice. As we have seen,
Hutton's geology had always stood outside the main channel of
geological theorizing. By the 1820s, his substantive claims had been
thoroughly discredited by the investigations of the Wernerians.
They had never adopted his theory of the consolidation of rocks;
they had transformed his theory of elevation almost out of recognition; and most tellingly, they had rendered totally implausible his
risky assumption that life on this planet had been indefinitely
maintained by their work in paleontology and stratigraphy. In
deciding to revive the Huttonian system, Lyell was quite out of line
with the generally Wernerian temper of British geology in the early
1820s.

   A possible reason for Lyell's choice is that, like Hutton, Lyell was
a deist. He believed that the deity's wisdom would be revealed in a
world designed to perpetuate a surface suitable for habitation by

plants, animals, and most particularly men. Lyell's deism may have predisposed him to admire the theory of the earth of his fellow Scotsman, Hutton. He appears to have been particularly appalled by Lamarck's suggestion that one species could be transformed into another (Bartholemew 1972).

But whatever his reasons, Lyell decided to resurrect the Huttonian system, particularly the thesis of the maintenance of habitability by the recycling of rocks, and bring it into line with the latest geological research. That in itself was a tall order. But perhaps not realizing how speculative the original Huttonian system (as opposed to Playfair's interpretation of it) had been, Lyell also wanted to develop a geological theory with impeccable methodological credentials. In Lyell's mind there was no better way to accomplish this than to adopt the method favored by Newton himself—the so-called *vera causa* method, or method of true causes—and adapt it to geology. Playfair had already represented Hutton as having taken the first steps in this direction. Lyell was one of many British scientists, including James Clerk Maxwell, who modeled their science on the methodologies advocated during the Scottish Enlightenment (Olson 1975). Lyell's particular genius was to adapt the *vera causa* method to the particular problems posed by geology.

Lyell may well have come across the *vera causa* principle in the works of Thomas Reid himself, the Scottish philosopher who had given the principle its canonical formulation (see chapter 1). He would certainly have encountered it in the works of two of his intellectual mentors, John Playfair and John Herschel (1792–1871). Playfair, in his treatment of the Huttonian theory, went to some pains to point out that the causes that Hutton was citing could be seen as *verae causae*. Herschel's example was no less important for Lyell. Son of the famous astronomer William Herschel, and a scientist with a growing reputation in his own right, Herschel's prestige was impressive (Cannon 1978).[1] Ever alert to factors that might affect his own reputation, Lyell had gone out of his way to persuade Herschel to write a companion piece to his own earliest publication, a paper on serpentine dykes in sandstone (Herschel 1825; Lyell 1825). Although Herschel's definitive treatment of the *vera causa* principle only appeared in his *Preliminary Discourse on the Study of Natural Philosophy* (1831/1966), when Lyell was already well embarked on the *Principles*, Herschel's general stand on this matter would have come as no surprise, but as welcome support to Lyell. As Herschel put it in the *Preliminary Discourse* (1830, 144),

> Experience having shown us the manner in which one phenome-
> non depends on another in a great variety of cases, we find
> ourselves provided, as science extends, with a continually increas-
> ing stock of such antecedent phenomena, or causes . . . , compe-
> tent, under different modifications, to the production of a great
> multitude of effects, besides those which originally led to a
> knowledge of them. To such causes Newton has applied the term
> *verae causae* that is, causes having a real existence in nature, and
> not being mere hypotheses or figments of the mind.

Later in the work, referring specifically to geology, Herschel (1830, 148) insisted that in explaining any phenomenon we must "seek, in the first instance, to refer it to some one or other of those real causes which experience has shown to exist, and to be efficacious in producing similar phenomena." Herschel used Lyell's theory of climatic change as his prime example of the effective use of *verae causae*.

Geologists who wanted to follow the *vera causa* method then, had to look for causes that were known to exist, though not necessarily in the situations they sought to understand, and they had further to look for evidence that the causes were adequate to produce the effects they sought to explain. However, Lyell introduced a new twist on the use of *vera causa*.

THE *VERA CAUSA* PRINCIPLE AND UNIFORMITARIAN GEOLOGY

Lyell's commitment to the *vera causa* method permeates the *Principles*. Yet in spite of the impressive bulk of the work, Lyell never discussed it explicitly, perhaps because the method was so thoroughly treated in philosophical works that he could assume that his audience would be completely familiar with it. However, we can reconstruct what he had in mind. In those many cases where the observational handicaps of the geologist were so great that he could not use the method of induction, what reasonable limits should the geologist put on the method of hypothesis? Lyell's answer was that all hypotheses about unobserved causes or effects must be founded squarely on what we have observed. The range of entities upon which geologists can draw to hypothesize about the unknown causes of a known effect are those agencies that have been observed in operation. Equally, in conjecturing about the unknown effects of some observed causes, they must similarly limit themselves.

Lyell's adherence to a *vera causa* method generated his principle of "uniformitarianism." The tenets of what William Whewell (1832), the Cambridge historian and philosopher of science and Professor of

Mineralogy, called Lyell's "uniformitarianism" were derived directly from the method of true causes. Historians of science have identified three distinct theses within Lyellian uniformitarianism (Hookyaas 1959; Rudwick 1970). The first of these, "law" uniformitarianism, asserts that the laws of nature have not changed over time; the second, "kind" uniformitarianism, that the kinds of geological causes have not changed over time; and the third, "degree" uniformitarianism, that the intensity of geological causes has not changed over time.

Law uniformitarianism was a relatively uncontroversial position in the early nineteenth century. Geologists agreed that the laws of nature neither changed nor had been suspended. The one possible exception was that "mystery of mysteries"—the origin of species. In the absence of a lawlike mechanism for the introduction of new species, many, perhaps most, geologists thought that new species were formed by the intervention of a creative power in the regular course of nature. Insofar as Lyell abandoned law uniformitarianism over the origin of species, he was no different from his contemporaries.[2] But with the exception of the origin of species, geologists were agreed that geological events resulted from the regular operation of nature's laws.

Kind and degree uniformitarianism were a different matter, the latter being particularly controversial. Claiming that the *kinds* of geological causes had never changed was bad enough. After all, certain causes at work today (ice, for example) simply could not have operated when the earth was—as many geologists believed with good reason—a hotter or even molten object. But claiming that the *intensity* of causes never changed (or, as Lyell sometimes put it, that the geologist could not reach back to investigate a period when forces were of a different intensity from those now observed) seemed no less than perverse. Most geologists believed that the evidence strongly suggested that the intensity of geological causes varied quite dramatically and was probably overall declining. Furthermore, if degree uniformitarianism was read as a stricture that geologists should restrict themselves to the period in which the intensity had not shifted, then it contracted the domain of the subject to no obviously good effect.

However, once we understand Lyell's commitment to the *vera causa* principle, puzzles about why he adopted kind and degree uniformitarianism vanish. Remember that Reid's first requirement for a true cause was that it "have a real existence, and not [be] . . . barely conjectured to exist." This was normally interpreted to mean that the cause in question should be observable; when adapted to geology it meant that geological causes must be observable. If we ask *which* geological causes are observable, the simplest answer is, those causes that are operating at present. And providing they *are* operating now, geologists can use them to explain effects at any period in earth history. Thus, if geologists wanted to invoke only "true" causes in geology, for a start, at least, they were limited to those causes that are observed to be operating at present. Equally, in conjecturing about the unknown effects of some observed cause, they must confine themselves in a similar way. In short, Lyell's requirement of kind uniformitarianism can be seen as a straightforward extension of the *vera causa* principle to a situation in which the cause and effect are widely separated in time.

Lyell derived degree uniformitarianism from the second plank of the *vera causa* principle—namely, that causes invoked "ought to be sufficient to produce the effect." Once again, when we ask how geologists were to know whether causes were *adequate* to produce the effects, the answer is that they should try to observe the cause-effect sequence operating. Thus, geologists had once again to limit themselves to presently operating causes. Only if they limited themselves to the intensity of causes observed at present could they be sure that the causes they invoked were adequate in degree to have caused the effects they sought to explain. And this, of course, is the requirement of degree uniformitarianism.

One might say that surely causes of *greater* intensity in the past would have been adequate to produce the effects, and thus Lyell ought to have been able to allow declining intensity of forces. This is to equivocate on the meaning of the word *adequate* in this context. A force of much greater intensity would not necessarily produce a similar effect; therefore, Lyell wanted to insist on identity of intensity.

However, the preceding discussion oversimplifies Lyell's position in an important way. Lyell did not simplistically equate "the present" and "the known." For him, the core dichotomy was between the observable and the unobservable, *not* between the present and the past. Lyell is often interpreted as being committed to "actualism," meaning that the past should be interpreted in terms of present causes (Rudwick 1970). If this were the extent of his position, then he would be inconsistent in arguing from past to present, as he often did—for example, claiming of volcanoes "that the points of eruption will hereafter vary, *because they have formerly done so*" (Lyell 1830–33, 1:313). However, this apparent inconsistency disappears if Lyell's basic commitment was to the *vera causa* principle, not to actualism.

The uniqueness of Lyell's methodological position can perhaps best be appreciated by comparing it with the superficially similar, but in fact different, methodologies adopted by George Poulett Scrope (1797–1876) and John Fleming (see chapter 7). Scholarly opinion has long asserted, even in the face of Lyell's own denials, that Lyell's methodology owes much to Scrope, since they were both "actualists." Rudwick (1974b, 229), for example, remarks that the parallels between their methodological comments "suggest that Lyell owed more to Scrope's *Considerations* than his public dismissal of that book would lead us to suppose." Besides many points of intersection in their careers (they shared the secretaryship of the Geological Society in 1825), Lyell and Scrope undoubtedly had many common intellectual concerns, including their skepticism about the popular elevation-crater and dolomitization theories. Like Lyell, Scrope worried that many geologists resorted to "sundry violent and extraordinary catastrophes, cataclysms, or general revolutions" to explain the changes that had taken place on the earth's surface. Such ill-formulated notions explained too much and too little, he thought, and had "the disadvantage of effectually stopping the advance of the science, by involving it in obscurity and confusion" (Scrope 1825, iv).

Determined to develop a better methodology, Scrope outlined "the only legitimate course of geological inquiry." If the geologist began "by examining the laws of nature which are actually in force, [he would] perceive that numerous physical phenomena are going on at this moment on the surface of the globe, by which various changes are produced in its constitution and external characters [that were] extremely analogous to those of earlier date, whose nature is the main object of geological inquiry" (Scrope 1825, v). Until such analogous causes had "been found wholly inadequate" to explain past changes on the earth, the geologist should not resort to other hypotheses. And in assessing whether they could explain the past, the geologist should make "the most liberal allowances for all possible variations and an unlimited series of ages." The key, claimed Scrope, was "Time!—Time!—Time!"

All this, Lyell (1827) heartily approved. For example, he followed Scrope's combination of the method of analogy and a liberal time allowance when explaining the origin of valleys in the Auvergne, and hence possibly all valleys.

Nonetheless, as Rudwick (1974b, 229–30) perceptively points out,

Lyell "could not accept the more speculative elements in Scrope's *Considerations.*" Lyell found Scrope's use of microtheories to explain volcanic action, for instance, much too permissive. Having no middle ground between the methods of analogy and hypothesis, as Lyell did in his *vera causa* requirement of independent observability, Scrope resorted to the method of hypothesis much earlier in the course of a geological investigation than Lyell thought warranted. Lyell used the method of hypothesis only when all else failed, preferring to take an agnostic stance rather than adopt a possibly unwarranted theory. Thus, Lyell's uniformitarianism owes more to his reflections on the merits of various extant methodologies than to Scrope, even though on many issues they adopted similar positions.

As to Fleming, he had argued that, since the warrant for the conclusion that past climates had changed was so shaky, the prudent geologist should assume that there had been no change (se chapter 6). This "uniformitarian" conclusion must have appealed to Lyell, since it agreed so well with his own Huttonian commitments. Yet, nonetheless, he disagreed with Fleming, since he believed that the evidence, imperfect as it was, indicated that the climate had changed. "As a staunch advocate for absolute uniformity in the order of Nature," Lyell wrote to Fleming (Lyell 1881, 1:260), "I have tried in all my travels to convince myself that the evidence was inconclusive, but in vain." In this case, Lyell's epistemological conscience overcame his repugnance for progressive changes. Much as he sympathized with Fleming's belief that, in the absence of any compelling evidence to the contrary, the geologist had to assume uniformity, Lyell thought that the evidence in this case *was* compelling. Fleming's uniformitarianism stemmed from his skepticism about the very possibility of constructing an empirically well-founded earth history. As such, its foundations and justification were quite different from Lyell's, which was based on a much more optimistic epistemological outlook.

In articulating the *vera causa* method for geology, Lyell had for the first time rooted the habit, common to geologists, of making inferences from the past to the present, and *vice versa*, in a respectable empiricist epistemology. Understood this way, Lyellian uniformitarianism is not a method developed solely to deal with the problems of geology, but an adaptation of a method widely advocated for use in *all* the sciences. Lyell himself was to use it to good effect in his theory of climate, though not in other parts of his geological system. The reaction to his theories was mediated by the contemporary understanding, since largely lost, of the structure of the argumentative strategies that Lyell was using. Most of Lyell's critics were less enamored of the *vera causa* method than he, finding it too restrictive.

LYELL'S THEORY OF THE EARTH

Lyell used his methodology to license a causal geology consistent with his Huttonian leanings, but radically unlike the theories advocated by von Buch, Humboldt, and Élie de Beaumont. He also investigated the historical geology of the Tertiary, but he was much less exclusively concerned with historical geology than most of his British contemporaries. Since good analyses of Lyell's theory are readily available (Rudwick 1970), I shall not describe it in detail. Instead, after giving a brief overview, I shall concentrate on his theory of climate, since that best exemplifies the strengths of his methodology, and on his theory of elevation, since that best exemplifies the limits of his methodology.

The question of habitability preoccupied Lyell, as it had Hutton. Like Hutton, Lyell rejected the belief that the forms of life on earth had changed progressively (Bowler 1976). In Hutton's day, this had been idiosyncratic but defensible; by Lyell's time, it was outrageous. Lyell conceded that the fossil record offered incontrovertible evidence of change at the level of individual *species*, but he denied that it warranted the more general conclusion that *classes* had changed. Instead, he argued that the same classes of plants and animals have always existed somewhere on the globe, although perhaps not always in the same location, since conditions at any spot on the globe constantly vary. Constancy of the classes of plants and animals was maintained through a succession of extinctions and creations of individual species (Lyell 1830–33, vol. 2; see also Hodge 1983).

Like his contemporaries, Lyell thought that species extinction was brought about by changing environmental conditions, particularly changing climate. And again like them, he found the introduction of new species rather more difficult to explain. Consistent with his methodology, he claimed that, since new species appeared to have been introduced regularly in the past, we should expect them to continue to be introduced regularly in the future. That no one had actually observed the introduction of a new species reflected how rarely the event occurred at any given spot on the globe. Speculation about its cause was fruitless since the geologist had no independent observational access to the cause.

What caused the changing environment that brought about extinctions was a difficult problem. Lyell granted the cogency of the paleontological evidence for climatic change, but he was unhappy with the increasingly popular explanation of climatic change in terms of a gradually cooling globe. Postulating a gradually cooling globe ran counter to Lyell's *vera causa* methodology by involving a cause that was completely removed from independent observation—namely, the interior heat of the earth. It had the further undesirable consequence that the earth would some day become uninhabitable. But this left him in a quandary. He was very unhappy with the theory that climate had deteriorated because the earth was cooling from a hot and molten body. This was a hypothetical speculation of just the kind that Lyell thought scientists should eschew, and it ran directly counter to his Huttonian belief in the continued habitability of the earth. A previously hot earth could not have supported life in the past, nor would it continue to do so if it cooled further in the future. Thus, Lyell had to find some alternative explanation for climatic change if he was to make good on his promise to construct a neo-Huttonian geology. At the eleventh hour, just as the *Principles* was about to be published, he saw a way of achieving this. Ironically, this last-minute effort to construct a theory of climate was to be his most successful use of the *vera causa* method.

LYELL'S *VERA CAUSA* THEORY OF CLIMATIC CHANGE

In the 1820s, Lyell had satisfied himself that the evidence for past climatic change was decisive (see chapter 7; see also Laudan 1982a). Thus, having accepted the evidence for climatic change, Lyell had to formulate a theory to account for this change in a manner that accorded with his methodological principles and his Huttonian convictions. In short, he had to "labour to account for the vicissitudes of climate, not to dispute them" (Lyell to Fleming, in Lyell 1881, 1:260).

A couple of weeks after this correspondence with Fleming, Lyell reported excitedly to his friend, the physician and geologist Gideon Mantell (1790–1857), that he had developed a new theory of climatic change:

> [I] am now engaged in finishing my grand new theory of climate, for which I had to consult northern travellers, Richardson, Sabine, and others, to read up Humboldt's and Daniel's last theories of meteorology, e& e&. I will not tell you how, till the book is out—

but without help from a comet, or any astronomical change, or any cooling down of the original red-hot nucleus, or any change of inclination of axis or central heat, or volcanic hot vapours and waters and other nostrums, but all easily and naturally. I will give you a receipt for growing tree ferns at the pole, or if it suits me, pines at the equator; walruses under the line, and crocodiles in the arctic circle. [15 February 1830, in Lyell 1881, 1:261–62)[3]

Lyell inserted his theory directly after the historical introduction to the *Principles*, before any other topic. It was scarcely a conventional opening for a geological treatise. They usually began with a discussion of mineralogy and petrology, proceeded through igneous and aqueous causes and then outlined historical geology. Depending on inferences from geological data, discussions of climate usually concluded—they did not introduce—geological books, if indeed they occurred at all. The prominence Lyell accorded climate bespeaks his concern with habitability and his pride in having constructed a *vera causa* theory.

Unlike earlier theories of climate that were based on the method of hypothesis, Lyell's theory incorporated a "true cause" or *vera causa*.[4] He inquired what existing and observable cause (*vera causa*) might be adequate to show how "such existing vicissitudes [of climate] can be reconciled with the existing order of nature" (Lyell 1830–33, 1:104). At one time, he had thought, "If a satisfactory explanation of so difficult a problem is ever obtained, we shall probably be indebted to astronomy for it" (Lyell 1826, 528), rejecting the possibility that geology might supply it. After all, the geologist's repertoire was limited to changes in physical geography, and these, Lyell believed, could at most redirect ocean currents. Thus the geologist was limited to "account[ing] for local fluctuations of climate in the same latitudes." Geological causes were far too "partial in their operation and limited in degree," to effect the major climatic changes that had occurred in the northern hemisphere. Just four years later, in his "grand new theory of climate," Lyell had changed his mind, arguing that changes in physical geography were *the* cause of global climatic changes. Astronomical, cosmogonical, and geophysical causes, like central heat, were not to be contemplated.

Lyell acknowledged that it was Humboldt's collection of climatic data that had given him the clue (Lyell 1830–33, vol. 1; Humboldt 1820–24, 3:14; see also Herschel 1832). Considering Humboldt's data, Lyell concluded that climatology, like geology, had been biased by unwarranted generalizations based on one small part of

the earth's surface. Just as geologists had assumed the Italian volcanoes to be typical of all volcanoes, meteorologists had assumed
that the European climate was typical of all climates. Humboldt had
shown this generalization to be too hasty. Geologists and meteorologists had not paid nearly enough attention to their peculiar position
as observers. Theories founded on observations of a small part of
the earth's surface were doubly misleading: first, the samples on
which they were based were not representative; and second, they
did not incorporate *geographical variation itself as a cause.*

Lyell found Humboldt's remark (1820, 3:14) that the "unequal
temperature of the two hemispheres . . . is less the effect of the
eccentricity of the earth's orbit, than of the unequal division of the
continents" particularly suggestive. If differences in physical geography caused differences in climate in the two hemispheres, why,
Lyell asked, going well beyond anything Humboldt had intended,
need the climatologist introduce changes in central heat or long-term
astronomical shifts? Maybe variations in physical geography could
explain all variations in climate, present *and* past. And physical
geography could be observed; central heat and long-term astronomical shifts could not, at least not directly.

Following Humboldt, Lyell (1830–33, 1:107) pointed out that
oceanic areas had "*insular* climates where the seasons are nearly
equalized," and continents had "*excessive* climates . . . where the
temperature of winter and summer is strongly contrasted." Lyell
used the distinction between oceanic and continental climates to
explain past climatic changes. One of the best-established facts of
geology, he asserted, was that the distribution of land and sea had
changed. The European strata showed that much of the Continent
had been under water in the relatively recent geological past;
observations of the erosive power of water showed that at this very
moment continents were being worn away and the materials for new
land masses deposited under the ocean. As the continents and
oceans gradually, but constantly, changed places, so the distribution
of the world's climates would change accordingly. Lyell now had a
cause for the present geographical variation of climate *and* a
potential *vera causa* for the past changes in climate. "Vicissitudes
of climate . . . must attend those endless variations in the geographical features of our planet," he concluded (Lyell 1830–33, 1:266–67).

Lyell carried out a thought experiment on possible changes in the
distribution of land and sea (1830–33, 1:112). He limited himself to
cases where (*a*) "the proportion of dry land to sea continues always
the same"; (*b*) "the volume of land rising above the level of the sea,

is a constant quantity; and not only that its mean, but that its extreme height, are only liable to trifling variation"; (*c*) "both the mean and extreme depth of the sea are equal at every epoch"; and (*d*) "the grouping together of the land in great continents is a necessary part of the economy of nature." Lyell did not believe that these conditions had always obtained, but was attempting to stay within "the strict limits of analogy." The first three conditions kept changes in the total area and volume of land and sea to the minimum. The fourth condition ensured that some continents remained. He speculated,

> It is possible that the laws which govern the subterranean forces, and which act simultaneously along certain lines, cannot but produce, at every epoch, continuous mountain-chains; so that the subdivision of the whole land into innumerable islands may be precluded.

Lyell now came to the point. Even these minimal shifts in the positions of land and sea could cause dramatic changes in climate because "whenever a greater extent of high land is collected in the polar regions, the cold will augment; and the same result will be produced when there shall be more sea between or near the tropics" (1830–33, 1:115). Lyell was not at all concerned that his critics might find his four conditions implausible, pointing out that his arguments would not be weakened but "greatly strengthened" if these larger variations were allowed (1830–33, 1:112).[5]

As an example, Lyell considered what would happen if Greenland were to sink into the Atlantic and a land bridge were to appear between Lapland and the Arctic island of Spitzbergen. The European climate would become more continental, the American climate more insular. Many European species would become extinct in the colder winters; many American species would become extinct in the milder climate. Yet this change was of the same order of magnitude as recent change in the Alps and Mediterranean.

If all the land masses of the Northern Hemisphere were to accumulate around the North Pole, greater effects yet would follow. It would be "difficult to exaggerate the amount to which the climate of the northern hemisphere would now be cooled down." Probably "no great disturbance can be brought about in the climate of a particular region, without immediately affecting all other latitudes, however remote" (Lyell 1830–33, 1:117). Circulating air and water currents would lower the mean temperature of the entire Northern Hemisphere. If all the land masses in the Northern Hemisphere were to accumulate round the equator, the climate of the whole hemisphere would become warmer. If, simultaneously, land in the Southern Hemisphere were to shift toward the Antarctic by a process of erosion, transport, consolidation, and elevation, then the mean temperature of the *entire* globe would decrease. Lyell did not pursue this, saying that not enough was known about the laws of elevation and subsidence to determine whether it could happen. Nonetheless, changing positions of land and sea, Lyell argued, were, in principle, adequate to change the climate of the globe.

From Lyell's point of view, his cause had the great mcrit of being reversible. Unlike the theory that the dissipation of the earth's interior heat caused climatic cooling, Lyell's theory did not entail that the climate would inevitably become colder and more excessive. Thus, Lyell could postulate continued habitability, for, instead of irreversible refrigeration, he could argue that the climate of each hemisphere, or possibly the whole globe, would swing between extremely cold and extremely warm periods. He described in detail how two extremes of climate—the "winter" and the "summer" of the "great year" or "geological cycle," as he called them—might be produced.

When Louis Agassiz (1807–1873) put forward the glacial theory in the late 1830s, Lyell had no difficulty assimilating it. In the next edition of the *Principles* (1847, 1:92), he summarized the history of climate, remarking that cooling had reached "its maximum of intensity in European latitudes during the glacial epoch, or the epoch immediately antecedent to that in which all the species now contemporary with man were in being." But the climate might well become warm again, and "the huge iguanodon might reappear in the woods, and the icthyosaur in the sea, while the pterodactyle might flit again through umbrageous groves of tree-ferns" (Lyell 1830–33, 1:123).

Lyell tried to establish that the known changes of land and sea had been of a type and magnitude to produce a gradually cooling climate. His evidence was limited to the Northern Hemisphere and the period since the oldest known fossiliferous strata (at that time thought to be Carboniferous) (Ospovat 1977). From the Carboniferous rocks, he concluded,

> The subaqueous character of the igneous products—the continuity of the calcareous strata over vast spaces—the marine nature of their organic remains—the basin-shaped disposition of the mechanical rocks—the absence of large fluviatile and of land quadrupeds—the non-existence of pure lacustrine strata—the insular character of the flora,—all concur with wonderful harmony to establish the prevalence throughout the northern hemisphere of a great ocean, interspersed with small isles. [Lyell 1830–33, 1:130][6]

Lyell tried to explain why, although he thought that the "laws which regulate subterranean forces are constant and uniform," the cooling of the globe had not proceeded steadily (1830–33, 1:139). In other words, he tried to show how uniform causes could produce nonuniform effects. In doing so, he was running counter to Cuvier's influential dictum that, since slow, regular causes could not produce sudden, catastrophic effects, there must have been "revolutions" in the past without present parallel (Cuvier 1817/1978). Lyell argued that, although geographical changes of the same magnitude always took the same time, their effects might be very different. A shift in land from one side of the equator to the other "would not affect the *general* temperature of the earth." A shift in land from low to high latitudes taking the same time would cause a marked "refrigeration of the mean temperature *in all latitudes*" (Lyell 1830–33, 1:140). This in turn could cause relatively rapid, large-scale extinctions of animals and plant life. In the most extreme circumstances, Lyell believed, "scarcely any of the pre-existing races of animals would survive." Therefore, the geologist ought not assume that species extinction occurred regularly over time, since changes in temperature "require unequal portions of time for [their] completion." The widespread extinctions between the deposition of the chalk and the Tertiary strata could have resulted from uniform causes, not from a catastrophic upheaval, as Cuvier and others had suggested.

Lyell castigated those geologists who had relied on hypotheses about central heat to account for climatic change. He attacked what was by then a straw-man combination of two theories—the cooling-earth hypothesis and the postulation of a primordial universal ocean. In a universal ocean, he said, the average depth of the water would have been much less, and the probability of ice much greater; hence, the postulation of a universal ocean was inconsistent with the cooling-earth hypothesis, with its consequence of a warmer earlier climate. He insisted that the existence of central heat should be proved and its changes thoroughly investigated before it was used as a panacea to explain all puzzling events in earth history. But, Lyell (1830–33, 1:86) lamented, it was probably impossible "to controvert, by reference to modern analogies, the conjectures of those who think they can ascend in retrospect to the origin of our system."

Lyell thought his theory of climate, founded on a *vera causa*, intrinsically superior to any set of "vague conjectures," a standard derogatory reference to the method of hypothesis. "[We] know that great changes in the external configuration of the earth's crust have at various times taken place, and we may affirm that they *must* have produced *some effect* on climate" (Lyell 1830–33, 1:143). His innovation was to argue that changing geography caused changing climate. Most geologists saw climatic change and geographical change as happening simultaneously, but independently; Lyell linked the two. He solved the problem of climatic change economically, demanding the minimum adjustment in the interpretation of existing knowledge.

John Herschel, one of the keenest advocates of the *vera causa* method, used Lyell's theory as his key example in the *Preliminary Discourse*. Unlike the theory of the overall cooling of the globe, which we do not "know" to have taken place, Herschel (1831, 47) felt that Lyell had offered "a cause on which a philosopher may consent to reason." However, he thought that Lyell had only established the first requirement for a true cause—namely, its existence—not the second, its adequacy. It remained to be seen "whether the changes actually going on are such as to warrant the whole extent of the conclusion, or are even taking place in the right direction." Consequently Herschel proposed a rival—namely, "the astronomical *fact* of the actual slow diminution of the eccentricity of the earth's orbit round the sun, affecting the *mean temperature of the whole globe.*"

Others were more cautious. The Royal Society excluded the theory of climate as a "controverted question" when awarding Lyell the Royal Medal for the *Principles* in 1834 (Cannon 1961). Whewell (1857/1961) praised its "ingenuity and plausibility" but thought that the problem involved "so many thermotical and atmological laws, operating under complex conditions" that he remained unconvinced. In short, the British agreed that Lyell's theory of climate was an ingenious, but in the last resort questionable, theory (Bartholemew 1979). Not until 1858, when Lyell was awarded the Copley Medal, did he receive institutional recognition for the theory.[7]

Lyell had constructed a theory of climatic change using the *vera causa* principle. But that theory rested on an assumption that land and sea had regularly and uniformly changed places. Thus, ultimately, Lyell's theory of climate relied on his theory of elevation.

LYELL'S THEORY OF ELEVATION

What Lyell needed, obviously, was a *vera causa* theory of elevation to support his *vera causa* theory of climate. Here he was much less successful. Volcanoes and earthquakes, which Lyell, like his contemporaries, thought were the primary causes of elevation and depression, were hard to investigate. Their causes were a matter for speculation and hypothesis, since, as Daubeny (1826, 356) had remarked a few years earlier, "the processes we undertake to consider are placed beyond the scope of natural observation, and can be conjectured solely from certain of their remote consequences." The method of hypothesis, Lyell thought, was unreliable; consequently, he eschewed hypotheses about the causes of volcanoes and earthquakes as far as he could. His lengthy chapters on the subject consisted almost entirely of descriptions drawn from his own tour of Italy and from historical testimony. His theory of elevation scarcely deserves the name, being embryonic in the extreme. It is best considered in relation to the theories being advanced by von Buch and Élie de Beaumont.

Lyell agreed with his continental contemporaries about many of the descriptive aspects of volcanoes and earthquakes. Volcanoes and earthquakes appeared to be localized along distinct lines. Their location appeared to have shifted in the past. Volcanoes appeared to be interconnected, since, when one was erupting, the others in the same area were dormant. Seawater appeared to play a part in eruptions, since so many volcanoes were found close to the sea. Volcanoes and earthquakes very probably had the same cause, since earthquakes were most severe where volcanoes were least explosive.

But for all these agreements, the differences are even more striking. Although volcanoes occurred along distinct lines, Lyell rejected the practice of dating mountain chains by their directions. He argued that the fossils contained in strata interbedded with igneous rocks offered the only reliable means of dating igneous rocks, and hence periods of uplift. Although earthquakes and volcanoes had shifted in the past, they had not been more common, as most geologists had assumed. The sum of volcanic and earthquake activity remained constant, Lyell claimed. The mineralogy of the igneous rocks occupied Lyell not at all, and he ignored the distinction between trachyte and basalt that was so crucial to the theory of elevation craters. Perhaps he was simply ignorant of mineralogy, or perhaps he thought it had been rendered redundant by the new paleontological methods. The intrusion of igneous rocks into strata was not a cause of elevation, as von Buch had claimed in the theory of elevation craters (see chapter 8). The quantity of volcanic products ejected annually, although enormous, was by no means enough to have formed the major mountain chains. Hence, volcanoes were not a major cause of elevation. Instead, earthquakes were the most effective causes of both elevation and depression.

At this point even Lyell was forced to resort to speculations about the causes of earthquakes and volcanoes, since a purely observational account left so much unanswered. Lyell hypothesized that the earth consisted of a molten interior overlaid by a floating solid crust containing cavities into which seawater leaked from time to time. Following the late eighteenth-century Newtonian John Michell (1724–1793), Lyell suggested that earthquakes were the result of waves in the molten interior (Michell 1760). Geysers occurred when seawater penetrating the cavities was converted into steam and forced to the surface. Volcanoes occurred when this steam forced molten lava up to the surface. Earthquakes elevated undisturbed segments of land, forming horizontal strata well above sea level. They also depressed segments, maintaining an earth of constant size in spite of the passage of minerals from the interior to the exterior by volcanoes.

Lyell tentatively suggested a number of causes of volcanoes and earthquakes, among them Davy's electrochemical theory of the oxidation of base metals, the venerable theory of the fermentation of pyrites and water, and a theory of the escape of gases liquefied under pressure. They were all, Herschel pointed out, "wanting in explicitness" (Herschel to Lyell, 20 February 1836, in Cannon 1961, 306).

Thus, Lyell had no well-articulated theory of elevation, and certainly no *vera causa* theory. Consequently, those who most admired Lyell's general program set about improving on his theory of elevation. Charles Darwin devoted all his young days as a geologist to modifying Lyell's theory in light of the much more sophisticated continental theories and his own observations (Darwin 1837a, 1837b, 1840, 1844/1897; Herbert 1970). John Herschel and Charles Babbage independently developed very similar *vera causa* alternatives to Lyell's theory (Greene 1983). Friends as undergraduates at Cambridge, Herschel and Babbage were agreed about the importance of true causes and quick to pick up clues about geographical causes from Lyell's theory of climate (Cannon 1978).

Casting around for true causes, Herschel and Babbage experimented with variations in physical geography, presumably prompted by the thought that Lyell's theory of climate might be extended to elevation. Herschel described how Lyell's account of "the Metamorphic Rocks'" encouraged him to "speculate how and why the mere fact of deep burial might tend to raise their temperature to the required point" (Herschel to Lyell, 12 June 1837, in Cannon 1961, 313). Herschel assumed that the earth consisted of a solid crust floating on a layer

> partly solid & partly fluid & composed of a mixture of fixed Rock, liquid lava, and other masses in various degrees of viscosity and mobility. [Assuming an equilibrium of temperature and pressure,] within the globe, the Isothermal Strata near the Center will be spherical, but where they approach the surface will by degrees conform themselves to the configuration of the surface of the *solid* portion, ie the bottom of the sea & the surface of continents. [Herschel to Lyell, 1836, in Cannon 1961, 306]

As material was eroded from the continents and deposited on the sea floor, the position of the isotherms, which depended on "the form of the bounding surface of the solid above," migrated upward. The overlying rock would become hotter, perhaps even melting, without any movement of molten material from the interior. In addition, because of the mixture of solid and liquid matter just below the crust, pressure differentials would be set up at the boundaries between continents and oceans.

Using this simple model, Herschel solved the "greatest difficulty in geology"—namely, "to find a *primum mobile* for the Volcano, taken as a general, not a local phenomenon," as well as a solution to the elevation of continents. Reviewing his theory for Lyell, he said:

> I don't know whether I have made clear to you my notions about the effects of the removal of matter from [continents] above to below the sea. 1st it produces mechanical subversion of the *equilibrium of pressure*—2dly it also, & by a different process . . . produces a subversion of the equilibrium of temperature. The last is the most important. It *must be an excessively slow process*. & it will depend 1st on the depth of matter deposited—2d on the quantity of water retained by it under the great squeeze it has got—3dly on the tenacity of the incumbent mass—whether the influx of caloric from below—which MUST TAKE PLACE acting on that water, shall either heave up the whole mass, as *a new continent*—or shall crack *it* & escape as a submarine volcano—or shall be suppressed until the mere weight of the continually accumulating mass breaks its lateral supports at or near the coast lines & opens there a chain of volcanoes. [Herschel in Cannon 1961, 310]

Thus, the *primum mobile* of the volcano (and of continental elevation) was simply

> the degrading power of the sea & rains (both originating in the suns action) above and the inexhaustible supply of heat from the

enormous reservoir below always escaping at the surface unless when repressed by an addition of fresh clothing at any particular part. [Herschel, in Cannon 1961, 310–11]

Now Herschel had a *vera causa* for elevation (Herschel to Murchison, 15 November 1836, in Babbage 1838/1967, 239–40). Once strata had been deposited, then "as a necessary consequence, and according to known, regular and calculable laws, heat *will* gradually invade them from below and around" (Herschel to Lyell, 1836, in Babbage 1838/1967, 240).

Like Lyell's *vera causa* for climate, Herschel's *vera causa* for elevation had an additional advantage for those, such as Lyell, who wanted to show that the earth was an essentially stable system. Since the "elevatory tendency is outwards," the globe would tend to "maintain its dimensions in spite of its expense of heat & thus preserve the uniformity of its rotation on its axis" (Herschel, in Cannon 1961, 311)

Like Herschel, Babbage speculated that the deposition of strata would cause the isotherms in the underlying deposits to shift upwards, assuming of course a reservoir of internal heat. Unlike Herschel, Babbage thought that the consequent *expansion* of the strata caused the elevation. Babbage calculated this expansion for various thicknesses of rock and increases in temperature, although these figures failed to convince Herschel. Unlike the refrigerating earth theory, but like Herschel's, Babbage's theory was founded on "really existing causes"—in this case the degradation of land masses and the deposition of eroded material on the sea floor, although the theory still did not satisfy both parts of the *vera causa* requirement. "The sufficiency of the theory for explaining all the phenomena cannot be admitted," he granted, "until it shall have been shown that their [the causes'] power is fully adequate to produce all the observed effects" (Babbage 1838/1967, 220).

LYELL'S IMPACT

The reception of the *Principles* has been much discussed (Hooykaas 1959; Cannon 1960, 1976; Rudwick 1967; Gillispie 1951/1959; Wilson 1972; Bartholemew 1976, 1979; Lawrence 1978). In Britain, reactions varied according to the philosophical viewpoint of the commentator. Herschel, who was himself an ardent advocate of the *vera causa* method, strongly supported Lyell's work. On receiving Lyell's gift of the fourth edition of the *Principles*, Herschel responded that he found the work "one of those productions which work a complete revolution in their subject by altering entirely the point of view in which it must thenceforward be contemplated" (Herschel to Lyell, 20 February 1836, in Cannon 1961, 305). He hoped that Lyell's "example [would] be followed in other sciences, of trying what *can* be done by existing causes, in place of giving way to the indolent weakness of a priori dogmatism—and as the basis of all further procedure enquiring what existing causes really are doing" (307–8). Whewell, who inclined to the view that the method was much too restrictive, and who was willing to endorse the use of hypotheses in science, was much less enthusiastic (Ruse 1976). Although he admired Lyell's philosophical acumen enough to devote two long reviews to the work, as well as a portion of his volumes on the *History* and *Philosophy of the Inductive Sciences*, he was critical, firmly placing himself and most of his contemporaries on the side he termed "catastrophist" in opposition to Lyell's "uniformitarianism" (Whewell 1831, 1832). Whewell found uniformitarianism, unlike the method of hypothesis, too restrictive. For the most part geologists were suspicious of Lyell's proposals for sweeping reforms. William Conybeare, joint author of the standard textbook on British geology and representative of the old guard of the Geological Society of London, united with Adam Sedgwick (1785–1873), Woodwardian Professor of Geology at Cambridge and the foremost geologist of the younger school. In his presidential address to the Geological Society in 1831, Sedgwick expressly compared Lyell's theory with Élie de Beaumont's and found the former wanting. Continental geologists read the French and German translations, appreciating Lyell's methodological ingenuity, but not changing their historical or causal theories.

But the fact of the matter is that, despite recent pioneering works on specific aspects of nineteenth-century geology by scholars such as Stephen Brush, Mott Greene, and Martin Rudwick, we still do not have even the outlines of an overall understanding of what happened during the rest of the century. Consequently, although what we do know tends to favor the position that Élie de Beaumont was more influential than Lyell, we simply have to wait for further historical scholarship to reveal the developments in geology after 1830.

10
Postscript

In chapter 1, I tipped my hand by expounding my views on the conceptual foundations of geology. In the balance of the text, I have argued that, until the last third of the eighteenth century, genetic causal theories dominated mineralogy and geology. The emergence of a strictly historical geology at the end of the century led to a separation between causal and historical geology. Although rarely in direct conflict and sometimes mutually supportive, geologists could pursue them largely independently. I have indicated the methods that geologists found most appropriate for achieving each of these two aims, stressing how self-aware they were about these methods. Hence, I have little to add to my discussion about the conceptual foundations of geology, except to point out that these are not closed issues. Historical geologists are still working at establishing the succession and correlating formations; they still debate the best way to do this and the reasons for the independence of formations. They are still filling in the details of the earth's past; they still debate the meaning and use of fossils. Similarly, geologists continue to investigate the causal processes shaping the face of the globe; they continue to be troubled by problems of access to the necessary data and the long time span between many geological causes and their effects. Of course, the theoretical apparatus they bring to bear is immeasurably better than it was in 1830, when my story ends, but that does not mean that foundational issues have vanished, only that they are continuing to be clarified as geology develops.

However, I do want to address three issues in this postscript that thus far I have mentioned only in passing, if at all. They are the relation of the history of geology I have told to other histories of geology, the function of the historiography of geology, and the relevance of this history to theories of scientific change.

By now, the reader familiar with the history of geology will realize that my history diverges significantly from the "received view" of the history of geology. As I read it, the received view consists of five interrelated theses: (1) geology took a historical turn when it became a "science" in the early nineteenth century; (2) the historical turn was accompanied (and probably caused) by a new empiricism, an enthusiasm for facts rather than theory, particularly facts gathered in the field; (3) both these developments were made possible by separating the earth sciences from religion, by distinguishing geology from Genesis; (4) they were coincident with the creation of institutions for the collection, publication, and transmission of geological knowledge; and (5) the making of modern geology was a British accomplishment.

The received view has been articulated in the British and American histories of geology published in the last half century. These have been dominated by two themes that follow from the received view. One is the heroic tale of the "discovery of time" and the re-creation of the earth's past (see, e.g., Gillispie 1951/1959; Haber 1959; Toulmin and Goodfield 1965; Burchfield 1975; Albritton 1980; Dean 1981; Rossi 1984). The other is the institutional and intellectual development of British geology in the eighteenth and nineteenth centuries (see, e.g., Gillispie 1951/1959; Porter 1977; Jordanova and Porter 1979; Rupke 1983b; Rudwick 1985). These histories, most of them based on meticulous scholarship, have not only taught us a great deal about their central topics but have also put us in a position to abandon many hoary old categories that have outlived their usefulness. For example, we no longer need to argue, as we did only a couple of decades ago, that categories like "neptunist," "vulcanist," "uniformitarian," and "catastrophist" confuse as much as they clarify in the history of geology.

But these kinds of revisionist moves have, by and large, been carried out within the bounds of the received view. Few of its core presuppositions have been challenged. Yet if the history I have told has anything to it, then some of these presuppositions do need to be rethought. To start, we have to uncouple the development of geology from the development of strictly historical geology. Then we can see that the latter did not displace but complement older causal and genetic traditions. Furthermore, we need to see that, from at least the seventeenth century, the issue for geologists was not whether they should be empiricists or theorists but the relation between theory and data, and between both and the aims of the geologist. We can then appreciate that fieldwork, for example, is not the only method, nor always the best method, of gathering geological data; it is a method for certain specific purposes. We need to recognize that geologists' religious attitudes were not monolithic. We can then understand why some geologists treated Scripture as testimony, why some denied its relevance, why some found that a deistic theology was heuristic in developing geological theory, and why others thought that Christian theology was detrimental to geology.

Once these kinds of moves have been made, we can rethink the classic problem of the received view—namely, when geology became a science. It is not a formulation of the problem that I am happy with, for reasons that I shall explain shortly. I believe that geology came of age in the late eighteenth century, not the early nineteenth, in the context of developments within mineralogy rather than in the context of a new interest in history and the outdoors, and in Germany and continental Europe rather than in Britain. Put in its bluntest form, I believe that the record shows that the Wernerians formulated the conceptual foundations of geology and dominated its intellectual and institutional development in the period between 1780 and 1830. But if this is so, why has it not been recognized?

I suspect that the prevailing historiography of geology owes much to historical accident, and even more to historical precedent. Historical accident because Charles Darwin started his career as a member of the British school of geology, and the success of evolutionary theory has cast a retrospective glamor over the geology of his mentors, particularly over Lyell's geology. This has meant that much history of geology has been written rather as a prelude to the theory of evolution than as a development in its own right. Historical precedent because Lyell composed a powerful apologia for much of the received view in the introduction to his *Principles of Geology* (1830–33), an interpretation that Archibald Geikie accepted and persuasively extended in his *Founders of Geology* (1905/1962). The latter is (rightly) still in print, remaining the most influential source on the history of geology for those who are not professional historians of science (and often for those who are). The influence of this historiography is, I believe, responsible for the received view.

Lyell's history establishes the received view more by what it does not say than by what it does. This is not the place to develop the historiographic argument in detail, but the outlines are clear. By attacking the effect of religion on geology, dismissing most cosmogony, by ignoring mineralogy, by treating Hutton and Werner as equally important, by mentioning only the Geological Society of London but no Continental institutions, and most of all by refusing to assess the contribution of any geologists subsequent to Hutton and Werner, Lyell by implication suggests the theses of the received view.

Geikie (1905/1962) wanted to describe how the "main foundations" of geology were laid. He owed the outlines of his history to Lyell, though he added the history of topics that had occupied him during his distinguished career as a geologist, including the role of heat and volcanoes, the importance of fieldwork in working out the succession, and an emphasis on geomorphology. The heroes of the story were the Huttonians, who had stressed the role of heat and the importance of geomorphology. Talking of Playfair's *Illustrations*, he said that he would impress on every student of geology "the advantage of reading and re-reading, and reading yet again this consummate masterpiece" (Geikie 1905/1962, 298). The most important lesson to be learnt from the history of geology, he said, was that "no branch of natural knowledge lies more invitingly open to every student who, loving the fresh face of Nature, is willing to train his faculty of observation in the field, and to discipline his mind by the patient correlation of facts and the fearless dissection of theories" (Geikie 1905/1962, 469).

Geikie's engaging history, then, reinforces the historiography of the received view that was initiated by Lyell. Just what a different story could be told at the time can be seen by comparing Lyell's history with Brocchi's history of Italian paleontology (1814), Cuvier's and d'Archiac's discussions of the history of geology (1810, 1847–60), or Friedrich Hoffmann's history of volcanic theory (1838), or by comparing Geikie's history with Zittel's (1901). These continental historians pick different periodizations, different central themes, and different heroes.

Of course, each of these historians was also a working geologist who expected his history to reflect and even to foster his own interpretation of the nature of geology. I do not say this critically, for that is the way I believe that history should be and, indeed, inevitably is. But I do say this to show that a necessary preliminary to telling the history of any subject is the careful examination of the historiography of the subject. Ignoring the historiography that shapes our perception of the past that we are trying to describe is as dangerous as ignoring the theories that shape our perception of the natural world that we are trying to investigate.

Finally, I want to summarize my thoughts, scattered through the volume, on what the development of geology tells us about the theories of scientific change put forward by scholars such as Thomas Kuhn, Imre Lakatos, and Larry Laudan.[1] I have been loathe to address this question. Historians of science tend to frown on colleagues who discuss the relevance of history to theories of change. They frequently express the suspicion that anyone familiar with these theories will inevitably shape their history in light of them, (a suspicion compounded in my case by my personal situation,) and hence that the resulting history is no test of the theories. Yet to refuse to talk about the relevance of my history to theories of science seemed somewhat cowardly, since I became a historian of science in the first place because I believed that "history, if viewed as a repository for more than anecdote or chronology, could produce a decisive transformation in the image of science by which we are now possessed" (Kuhn 1962, 1).

Rather than try to find which theory of change best fits the historical record, I have elected to examine some of the major commonalities between the theories, as well as to comment on a couple of the differences.We now know that, despite the many divergences between the major theories of change, there is a substantial degree of consensus among them (Laudan et al., 1986). Moreover, this consensus often runs counter to the commonsense picture of science espoused by the positivists a generation ago and by most of the general public to this day.

The claims held in common by theorists of scientific change that I shall consider are the following: that the most important units for understanding scientific change are large-scale, relatively long-lived conceptual structures, variously referred to as "paradigms," "global theories," "research programmes," or "research traditions," which we can group under the term *guiding assumptions*; that, contrary to Popper, refutations are not sufficient by themselves to overthrow guiding assumptions; that data do not fully determine choice of theories or guiding assumptions, but that theories in other parts of science, as well as metaphysical, theological, and other nonscientific factors play a role; that guiding assumptions play an important heuristic role and are in part selected in light of their promise in that regard; and that, in spite of the overall growth of knowledge, there are losses as well as gains during scientific development.

That there are guiding assumptions in the history of geology is quite clear. I have already discussed the matter at some length in chapter 5 with respect to the Wernerian radiation. Trying to tell the development of geology on the scale adopted in this book without terms like the *Becher-Stahl cosmogonic tradition* or the *Wernerian radiation* is not just difficult; it is unfaithful to the historical record.

The scientists I have studied placed their theories quite self-consciously in the context of guiding assumptions. The fact that not every scientist subscribed to a set of guiding assumptions, or that those who did had differences of opinion, does not mean that most scientists cannot be placed within one or another set of guiding assumptions current in a given period.

The significance of refutations is somewhat more difficult to assess. Most sets of guiding assumptions in the history of geology were sufficiently loosely articulated that no single piece of evidence could decisively discredit them. Parts of a theory that proved to be inadequate could always be patched up. But where more specific theories were concerned, refutations did count. The most striking example is Hutton's theory of consolidation. As I argued in chapter 6, most geologists thought that the behavior of calcareous earths on heating was sufficient reason to reject the theory. And this particular theory was so central to Hutton's general program that the rejection of the theory threw doubt on the wider set of assumptions. Therefore, although straightforward refutations may not occur, negative evidence does play a more important role than many of the theorists of change have been willing to allow.

On the question of whether data from the field in question alone is enough to determine whether or not a theory is accepted or rejected, I am of one mind with the theorists of scientific change in thinking that it is not. Much of this book is a sustained argument to the effect that the development of geology can only be understood in the light of interaction with other disciplines such as chemistry, systematics, and, to a lesser extent, physics. Geological theories were judged in light of their consistency with theories in other disciplines as well as in terms of their relation to purely geological data. I have not paid so much attention to extrascientific factors, but they do appear to have been important in the reactions to, say, the theories of Hutton and Werner.

I also agree with the theorists of scientific change that guiding assumptions have an important heuristic function and that they are assessed, in part, on the basis of that function. I believe that one reason that geologists were so excited about Werner's theory is that he laid out a clear program of research, so that both historical and causal geologists knew how to modify his theories when they ran into difficulty. Hutton's theory was not nearly so successful heuristically. He himself discouraged the laboratory experiments that were the most obvious way to improve his theory of consolidation. He suggested no way to make this theory of elevation more precise or more testable.

And finally, I think that the theorists of scientific change are correct in saying that there are sometimes losses as well as gains in the course of scientific development. I have not found many examples in my particular historical episode, but one case does stand out in my mind. That is Werner's loss of the ability to explain the order in which the various rocks of the earth's crust had been formed (see chapter 6). This is a real and significant problem. Geologists once had an answer couched in terms of the differential deposition of minerals from the ocean. As they came to see that not all rocks had been deposited from water, they simultaneously found that they could not even explain why the primary rocks differed from the secondary in such systematic ways, let alone the sequence within these major classes.

The history of geology seems to me to offer clear guidance in two cases of significant disagreement among theorists of scientific change. One is over the nature of guiding assumptions and the way they change. Although the theorists of scientific change usually consider guiding assumptions to contain statements of aims, methods, and goals, they are divided over the issue of whether these elements are locked together, so that they all change together in a revolutionary upheaval, or whether the elements can be changed one at a time, allowing for more gradual, piecemeal change. For reasons that I explained in chapter 5, I believe the latter option much more accurately represents the historical record. The other case of disagreement has to do with whether immature and mature science differ significantly, or, in terms that historians are more apt to use, whether it makes sense to talk of the founding of a science. Here again I am a gradualist. Although geology in 1830 differed greatly from geology in 1730, I do not think there was one point where the introduction of a new "paradigm" turned it from an immature science to a mature science in a matter of a year or so. Like most historians of science who have examined in detail episodes of apparently rapid change, I am struck by the continuities rather than the sharp breaks or revolutions. Werner's theory marks a turning point between the old Becher-Stahl cosmogony and the Wernerian radiation, rather than a revolution. The conceptual foundations of geology were not the master stroke of one man, or even of a small number of men, but were hammered out over half a century or more through the disputes and discussions of those who concerned themselves with the structure, the processes, and the past of the earth.

Notes

CHAPTER ONE

1. Since historians of science have become so sensitive to whiffs of "Whiggish" anachronism revealed by the anachronistic use of terms, I need to say a word about terminology. Where possible I shall use the word that contemporaries preferred. Thus I shall refer to *cosmogony* and *mineralogy* in the seventeenth and eighteenth centuries and *geology* in the last third of the eighteenth century and the nineteenth century (Adams 1932; Dean 1979). Although *geognosy* was commonly used in Germany and France, I shall usually substitute *geology*. *Geognosy* never attained currency in English and, in any case, meant much the same as *geology*. When the use of *mineralogy* or *cosmogony* might be obscure to the modern reader, or when I wish to refer to the spectrum of activities that roughly corresponds to the modern use of geology, I shall use *geology*. I shall also use the terms *historical* and *causal* geology to distinguish two distinct activities in the early century because even though they are, strictly speaking, anachronistic there were no common equivalents at the time. To refuse to allow this historical privilege is to be obscurantist and pedantic.

Indeed, to my mind, altogether too much has been made of changes in terminology. The fact that *geology* came into currency in English during the early nineteenth century is often taken as evidence that the discipline "emerged" only then. The term, it is claimed, should not be used in discussions of earlier periods since before its introduction "geology did not exist." That conclusion is, to say the least, questionable. And attempts to substitute other terms lead to clumsy circumstances or equally anachronistic choices. For example, some authors have resorted to "the terms 'earth sciences' or 'environmental sciences' [as] convenient alternatives since they are loose concepts which merely denote the objects of study" (Jordanova and Porter 1979, xi). Since the terms *earth science* and *environmental science* were invented in the 1960s in order to reformulate the boundaries of geology so as to accommodate new sources of funding associated with the space race and oceanographic exploration and new sensibilities associated with the environmental movement, they are no more neutral or timeless than the older term, *geology*.

2. For a parallel shift within biology, consider the comment by Sloan (1979, 117–18) that "the critical innovation in Buffon's species concept [is that] the very reality of the species was a *temporal* reality."

CHAPTER TWO

1. Porter (1977, 56) overstates his case when he claims, "So long as minerals and metals were believed generated by unique chemical combinations, stoked by local fires or chemical juices, or activated by special crystallizing tendencies, on the analogy of petrifying waters, mineral studies were not likely to spill over into an interest in structure or strata [where the latter are construed to be the crucial entities for geology as opposed to mineralogy.]" This and the following three chapters will be devoted to showing just how crucial mineralogy was to the development of geology.

2. Indeed, the doctrine persisted well into the nineteenth century. See, for example, Phillips (1816, 15, 24, 29, 52).

3. Aristotle's *Meteorologica* reappeared in the West in the eleventh and twelfth centuries, and generally included a section entitled "De Mineralibus." Since this section contained Arabic proper names, there was speculation that this was not part of the Aristotelian corpus, but an Arab insertion. In the first half of the thirteenth century, Albertus Magnus correctly recognized it as Avicenna's work, and he referred to it frequently in his *Book of Minerals* (Wyckoff 1967; also see Holmyard and Mandeville 1926). Most other Greek works on mineralogy had been lost. Of the twenty books on mineralogy that Pliny reported he consulted before compiling his *Natural History* in the first century A.D., all but Theophrastus's brief essay "On Stones" vanished. Pliny's own encyclopedic but disorganized discussion of some six hundred different minerals significantly increased the available information about minerals, but it was of no immediate consequence for a theory of the earth.

4. See Carozzi and Carozzi (1984, 285–87) for some eighteenth-century equivalents of these terms.

5. This is a little puzzling, though Eichholz has suggested that the dry exhalation may have been the efficient, rather than the material, cause of these minerals (Eichholz 1949).

6. Boyle published his main mineralogical work, the *Essay about the Origine and Virtues of Gems* in 1672. He intended it as a portion of a more lengthy discussion of minerals that never appeared, and its contents are of much more general significance than the title indicates. The leading exponent of the corpuscular philosophy in England, Boyle delighted in showing how chemical phenomena could be explained in terms of matter in motion. He intended to turn chemistry into a full-fledged branch of natural philosophy. Since minerals were some of the most important

substances with which chemists worked, they fell under this general program. Boyle had considerable practical experience with minerals. He was a member of the Company of Mines Royal, which, like its sister company, the Company of Mineral and Battery Works, had been founded by royal charter in 1568 as a way of providing the capital for expensive mineral workings, particularly the copper mines in Cumberland (Hall 1965). As a member of the company, Boyle regularly inspected mines. He had his own mineral cabinet, which he bequeathed to the Royal Society, and the Society chose him to assemble their questionnaire on mines and mining.

7. Rudwick (1972) gives an excellent account of the history of paleontology, though his decision to leave out the eighteenth century means that there is still no acceptable account of paleontology in that period.

8. Some scholars, particularly in the British Isles, were still making similar arguments at the end of the eighteenth century (e.g., Kirwan 1799).

9. A slim octavo volume of about 75 pages, the *Prodromus*, as the title suggests, like Boyle's essay on gems, was intended as an introduction to a longer piece that never appeared.

Steno's mineralogical investigations had been set in train by the landing of a large shark near the Italian town of Livorno in October 1666 (Rudwick 1972). The Grand Duke of Tuscany, Ferdinand II, in whose realm this startling event had occurred, requested that Steno dissect the beast. Steno was a natural choice, as an accomplished physician and a member of the Accademia del Cimento, founded by Leopold de' Medici just a few years earlier, in 1657, to provide a forum for the discussion of natural philosophy. Steno was particularly intrigued by the shark's rows of triangular teeth. They looked identical to the "tongue stones" (or glossopetrae) found on the island of Malta, the origin of which had puzzled generations of scholars. Why, asked Steno, should there be a similarity between bodies that grow in the sea, such as shark's teeth, and bodies found on land, such as "tongue stones"? Indeed, how did solid bodies grow from fluids such as air and water? The *Prodromus* was an attempt to answer such questions. Since minerals had presumably grown this way, Steno (1669/1968, subtitle) thought that his answer would also provide a "foundation for the rendering a rational accompt both of the Frame and the several changes of the masse of the earth, as also of the various productions therein."

Henry Oldenburg, the secretary of the newly formed Royal Society of London, translated the *Prodromus* into English in 1671, only two years after its original publication. In his introduction, he heralded it as an inquiry into "the true knowledge of one of the Great Masses of the World, the *Earth*." As such, it contributed to a general program to explain the earth in terms of the mechanical philosophy and was an apt companion to the attempt by Boyle (1673, introduction) to "lead us on a great way in the knowledge of another of the great masses, the AIR." Shortly thereafter, Steno's monograph was published as an appendix to a collection of Boyle's essays that included the *Origine and Virtues of Gems* (1673).

10. It is tempting to argue that Becher was reacting to the physical cosmogonists, though I can find no textual evidence for this. But, as the title and the dedication indicate, he was responding to Athanasius Kircher's *Mundus subterraneus* (1665), a work that fell squarely within the organic tradition.

CHAPTER THREE

1. There were a few notable exceptions to the disdain shown by academicians for practical mineralogy. Leibniz, for example, took a keen interest in mining.

2. A useful account of at least part of the story, that dealing with Bertrand and Lehmann, can be found in Carozzi and Carozzi (1984).

3. The great Dutch chemist Hermann Boerhaave was almost alone in resisting Stahlian chemistry. We shall examine his impact on geology in chapter 6.

4. In England, only a few individuals, such as John Hutchinson (1674–1737) and Alexander Catcott (1725–1779), continued to try to produce physical cosmognies (Porter 1977).

CHAPTER FOUR

1. Later in life, Linnaeus developed a theory of species generation that, had it been generally recognized, would have lent itself quite well to minerals. According to this, God initially created a few elemental species, and all other species arose by hybridization. Hence, most species were "molecular" (David Hull, letter, June 1985). But this was not the part of Linnaean theory that mineralogists adopted.

CHAPTER FIVE

1. Of course, Werner was not the only geologist to advocate fieldwork. Many others, including Bertrand and Bourguet in France, did so. Werner's contribution was to make a compelling intellectual rationale for fieldwork. See Davies (1969, 146) for a similar argument.

2. Recent research has shown the interaction between the German philosophy of the period and physiology, and it seems that the same was true of mineralogy and geognosy (LeNoir 1982).

CHAPTER SIX

1. Various other sources for Hutton's commitment to cycles have been suggested, ranging from his experience with cyclic physiological systems as a medical student, to his contact with natural cycles as a farmer, to his perception of parallels with the steam engine designed by James Watt, an

associate of Hutton's good friend Joseph Black (Davies 1969; Donovan and Prentiss 1980). Probably all these influences played a role.

2. Playfair's interpretation of Hutton was adopted by most later commentators, notably Charles Lyell, Archibald Geikie, and E. B. Bailey. It continues to flourish in the popular literature on the history of geology (Lyell 1830–33; Geikie 1905/1962; Bailey 1967; McPhee 1980).

3. The standard account has Hall as the pioneer of experimental geology (Geikie 1905/1962; Eyles 1961). But experiments were the norm in eighteenth-century geology, and hence Hall is best seen as a traditionalist, reverting to eighteenth-century experimental techniques in mineralogy and marking the continuation or even the demise of an old tradition, rather than the inauguration of a new one.

CHAPTER SEVEN

1. For a fascinating account of the routine, though not simple, work of the historical geologist as it developed in the 1840s, see Rudwick (1985).

2. Rhoda Rappaport (1982) has drawn attention to the use of the term *accidental* in an interesting paper. She locates the term within the vocabulary of the general historian, arguing that eighteenth-century writers used *accident* to indicate purely local features, or deviations from the normal and the natural. This is not, I think, at odds with my account, which looks at its place in the technical vocabulary of natural history.

3. I am indebted to the Peter Connor of the Classics Department of the University of Melbourne and Glen Bugh, Virginia Tech for useful discussion of this point.

4. I have always wondered why the discovery of the extraordinary fauna of Australia in the late eighteenth century did not give Cuvier and his contemporaries pause.

5. One of the earliest maps to show the areal distribution of the rocks was constructed by Georg Füchsel (1722–1773) to accompany his mineral geography of Thuringia (1761). It was published in an obscure regional journal, the *Acta* of the Erfurt Academy of Sciences, in a very poor Latin translation, and appears to have had limited impact, though it is possible that Werner was indebted to Füchsel for some of his ideas. In particular, Füchsel defined a formation as rocks formed at the same time, of the same material, and in the same manner, a definition clearly close to Werner's. I am grateful to Bert Hansen for supplying me with a copy of his translation of Füchsel's work.

6. I am grateful to Jim Secord for this reference.

7. In his *Époques de la Nature*, Buffon (1778) had already made the inference from extinct species. He noted the giant size of the ammonites and other fossils found in the secondary strata; since he knew that most species grow to larger sizes in the tropics, he argued that a tropical climate had prevailed when those formations were deposited, even in what are now temperate latitudes (see also Rudwick 1972). The bones of

elephants discovered in North America, northern Europe, and Siberia offered independent evidence for this inference. Buffon believed that they showed that, relatively late in the earth's history, tropical mammals had flourished as far north as zones that are now subarctic.

8. Conybeare (1829, 145–46) remarked sarcastically, "The narrow system of Oxford logic in which I have unfortunately been trained, renders me less sensible to all the merit of such modes of ratiocination." Referring to the diluvial hypothesis advanced by Conybeare's Oxford colleague, William Buckland (1784–1856), Fleming (1829b, 68) jeered, "With the exception of a very few individuals who may still be found on stilts, amidst the 'retiring waters', the opponents of the hypothesis have become as numerous as were formerly its supporters, and the period is probably not far distant, when the *Reliquiae diluvianae* of the Oxonian geologist will be quoted as an example of the *idola specus*." On reading this, Lyell cautioned Fleming, and urged him to "keep in well with some party" or a recent nickname, the "Zoological Ishmael," might stick (Lyell to Fleming, 3 February 1830, in Lyell 1881, 1:260).

CHAPTER EIGHT

1. There is no reason to treat this report with suspicion. Fossils are still found associated with interbedded basalt flows for example, at Portrush in Northern Ireland.

2. Later commentators, such as Suess (1904–9) assumed that von Buch believed that the area was being elevated by igneous forces, or, to use von Buch's later terminology, that it was a nascent elevation crater. It is hard to find evidence for this in von Buch's own writings.

CHAPTER NINE

1. David Hull found, in a survey of four scientific journals for the year 1867, that Herschel was cited more frequently than any other scientist (Hull, letter, June 1985).

2. David Hull has suggested in a private communication that Lyell may have been unwilling to contemplate the possibility that species evolved because he knew that this would carry the consequence that they were no longer natural kinds and could not function in natural laws.

3. Edward Sabine (1788–1883) had been the official astronomer on arctic expeditions in 1818 and 1819. John Richardson (1787–1865) was surgeon and naturalist on arctic expeditions in 1818–22 and 1825–27. Both recorded and published observations of the climate. John Daniell (1790–1845) had codified contemporary meteorological knowledge in his *Meterological Essays* (1823).

4. Although I agree with Dov Ospovat that Lyell's theory of climate occupied a central role in his geological system, I disagree that "Lyell's version of the principle of uniformity resulted from his desire to justify a highly speculative interpretation of the evidence about climate, an

interpretation whose purpose was to preserve man's unique status in creation'' (Ospovat 1977, 318).

5. In other parts of his geological theory, such as those dealing with the intensity and frequency of mountain-building activity, Lyell chose *not* to argue that regular changes in the interior of the globe could, by similar threshold effects, have led to intermittent periods of intense mountain building on the exterior, as Élie de Beaumont was to suggest at about this time. Lyell probably treated these two cases differently because of the different weight he gave to the evidence for the respective changes to be explained. In the case of change of species (and the changes in climate that he believed had caused this), Lyell was convinced that the evidence supported the claim that the changes had been spasmodic. In the case of mountain building, Lyell was far from convinced that the evidence warranted the assertion of sudden bursts of activity.

6. Lyell's conclusion that the Northern Hemisphere was oceanic in the Carboniferous might seem to be at odds with Lyell's fourth condition for the redistribution of land and sea—namely, that continents were always a part of the economy of nature. But as long as there were continents *somewhere* on the earth's surface—for example, in the Southern Hemisphere—the condition was, *in fact*, fulfilled. In any case, as we have seen, Lyell thought that actual topographical variations might have been much greater those allowed in his thought experiment.

7. Darwin's *Origin of Species* nearly suffered a similar fate when Sabine recommended that it not be mentioned in the list of works for which Darwin was awarded the medal. I am indebted to David Hull for this information.

CHAPTER TEN

1. I have largely ignored the claims of recent sociologists of science. Their claims seem to wobble perilously between the implausible and the innocuous. If they are claiming that the world does nothing to shape scientists' beliefs, and the social and psychological situation does everything, then the claim is implausible (not to mention landing the historian in a reflexive vortex from which no escape seems possible). If they are claiming that the world partially shapes scientists' beliefs and the social and psychological situation also plays a role, then the claim seems true but innocuous. To have some teeth, some theses about how the world, the social situation, and the psychological situation interact to cause belief will have to be advanced.

Bibliography

MANUSCRIPTS

Banks, Joseph. Papers. British Library.
Buch, Leopold von. Papers. Berlin Geology Museum.
Greenough, George Bellas. Papers. University College, London.
Greenough, George Bellas. Papers: Map Collection, Geological Society of
    London. University Library, Cambridge.
Saussure, Horace Bénédict de. Papers: Scientific Correspondence of Swiss
    Geologists. University Library, Geneva.
Smith, William. Papers. Geology Museum, Oxford University.

PRINTED SOURCES

I have made extensive use of the *Dictionary of Scientific Biography*. Rather than list all the entries I have consulted for biographical information, I have trusted that the reader will take this as given.

Adams, Frank D. 1932. Earliest use of the term geology. *Bulletin of the
    Geological Society of America.* 43:121–23.
———. 1934. Origin and nature of ore deposits: An historical study. *Bulletin
    of the Geological Society of America.* 45:375–424.
———. [1938] 1954. *The birth and development of the geological sciences.*
    Reprint. New York: Dover.
Agassiz, Louis J. R. 1848. *Bibliographia zoologiae et geologiae: A general
    catalogue of all books, tracts and memoirs on zoology and geology.*
    London: Ray Society.
Agricola, Georgius. [1546] 1955. *De Natura Fossilium* (Special Paper 63).
    Trans. M. C. Bandy and J. A. Bandy. New York: Geological Society of
    America.
Aguillon, Louis. 1889. L'École des Mines de Paris, notice historique.
    Annales de Mines. 8th series, 15:433–686.
Aikin, Arthur. 1797. *Journal of a tour through North Wales and part of
    Shropshire.* London.

Albritton, Claude C. 1963. *The fabric of geology.* Reading, Mass., and London: Addison-Wesley.

———. 1980. *The abyss of time: Changing conceptions of the earth's antiquity after the sixteenth century.* San Francisco: Freeman, Cooper and Co.

Albury, William R., and D. R. Oldroyd. 1977. From Renaissance mineral studies to historical geology, in the light of Michel Foucault's "The order of things." *British Journal for the History of Science.* 10:187–215.

Allan, D. E. 1976. *The naturalist in Britain.* London: Allen Lane.

Allen, Don Cameron. 1963. *The legend of Noah.* Urbana: Univ. of Illinois Press.

Archiac, Adolphe d'. 1847–60. *Histoire des progrès de la géologie de 1834 à 1845.* 8 vols. Paris: Sociéte Géologique de France.

Arduino, Giovanni. 1782. *Memoria epistolare sopra varie produzioni vulcaniche, minerali e fossili.* Venice: B. Milocco.

Aristotle. 1952. *Meteorologica,* translated H. D. P. Lee. Loeb Classical Library. London: Heinemann

———. 1950. *De Generatione et Corruptione.* Translated E. S. Forster. Loeb Classical Library. London: Heinemann

Arkell, W. J., and S. I. Tomkieff. 1953. *English rock terms, chiefly as used by miners and quarrymen.* Oxford: Oxford Univ. Press.

Artz, F. B. 1966. *The development of technical education in France, 1500–1800.* Cambridge, Mass.: MIT Press.

Ashworth, William. 1970. Backwardness, discontinuity, and industrial development. *Economic History Review.* 23:163–69.

Ashworth, William B. 1984. *Theories of the earth, 1644–1830: The history of a genre.* Kansas City, Mo.: Linda Hall Library.

Aubuisson de Voisins, J. F. d'. 1819. *Traité de géognosie, ou, exposé des connaissances actuelles sur la constitution physique et minérale du globe terrestre.* 2 vols. Strasbourg and Paris.

Babbage, Charles. 1833. Observations on the temple of Serapis. *Proceedings of the Geological Society of London.* 2:72–76.

———. [1837] 1967. *The ninth Bridgewater treatise.* Reprint of 2d ed., 1838. London: Frank Cass.

Bailey, E. B. 1962. *Sir Charles Lyell.* London: Thomas Nelson.

———. 1967. *James Hutton: The founder of modern geology.* Amsterdam: Elsevier.

Baker, Keith M. 1975. *Condorcet: From natural philosophy to social mathematics.* Chicago: Univ. of Chicago Press.

Bakewell, Robert. 1813. *An introduction to geology.* London.

Barlow, Norma, ed. 1958. *The autobiography of Charles Darwin, 1809–1882.* London: Collins.

Bartholomew, Michael. 1972. Lyell and evolution: an account of Lyell's response to the prospect for an evolutionary ancestry for man. *British Journal for the History of Science.* 6:261–303.

————. 1976. The non-progress of non-progression: Two responses to Lyell's doctrine. *British Journal for the History of Science.* 9:166–74.

————. 1979. The singularity of Lyell. *History of Science.* 17:276–93.

Baumgärtel, Hans. 1959. Alexander von Humboldt und der Bergbau. *Freiberger Forschungshefte.* D33. Berlin: Akademie-Verlag.

————. 1961. *Aus der Geschichte der Bergakademie Freiberg.* 3d ed. Berlin: Akademie-Verlag.

————. 1963. Bergbau und Absolutismus: Der sächsische Bergbau in der zweiten Hälfte des 18. Jahrhunderts und Massnahmen zu seiner Verbesserung nach dem Siebenjährigen Kriege. *Freiberger Forschungshefte: Kultur und Technik.* D44:80–81.

Becher, J. J. 1668. *Actorum laboratorii chymici monacensis seu physica subterraneae.* Frankfurt: Zunneri.

Beddoes, Thomas. 1791. On the affinity between basaltes and granite. *Philosophical Transactions of the Royal Society of London.* 81:48.

Beer, Sir G. de. 1962. The volcanoes of the Auvergne. *Annals of Science.* 18:49–61.

Berger, J. F. 1811. A sketch of the geology of some parts of Hampshire and Dorsetshire. *Transactions of the Geological Society of London.* 1:249–68.

Bergman, Torbern Olaf. 1783. *Outlines of mineralogy.* Trans. W. Withering. Birmingham: T. Cadell.

Beringer, C. 1954. *Geschichte der Geologie und des Geologischen Weltbildes.* Stuttgart: Enke-Verlag.

Berman, Morris. 1975. 'Hegemony' and the amateur tradition in British science. *Journal of Social History.* 9:30–50.

Bertrand, Élie. 1752. *Mémoires sur la structure intérieure de la terre.* Zurich: Heidegger.

————. 1763. *Dictionnaire universel des fossiles propres et des fossiles accidentels, contenant une description des terres, des sables, des sels, des soufres, des bitumes.* Avignon: L. Chambeau.

Birembaut, Arthur. 1953. Les préoccupations des minéralogistes français en 18è siècle. *La 72è session de L'Association Française pour L'Avancement des Sciences.* 534–38.

————. 1964. L'enseignement de la minéralogie et des techniques minières. In *Enseignement et diffusion des sciences en France an XVIII siècle,* ed. R. Taton, 372–76. London: W. Blackwood.

Black, Joseph. [1755] 1902. Experiments upon magnesia alba, quicklime, and some other alcaline substances. Reprint. Chicago: Univ. of Chicago Press.

————. 1794. An analysis of the waters of some hot springs in Iceland. *Transactions of the Royal Society of Edinburgh.* 3:95–126.

————. 1803. *Lectures on the elements of chemistry.* Ed. John Robison. 2 vols. Edinburgh.

Blake, Ralph M., Curt Ducasse, and Edward Madden. 1960. *Theories of scientific method: The Renaissance through the nineteenth century.* Seattle: University of Washington Press.

Bloor, David. 1982. Durkheim and Mauss revisited: Classification and the sociology of knowledge. *Studies in History and Philosophy of Science.* 13:267–97.

Blumenbach, J. F. [1806–11] 1865. *The anthropological treatises.* Trans. Thomas Bendyshe. London: Longman.

Boerhaave, Hermann. 1735. *Elements of chemistry.* Trans. from 1732 Latin edition. London: J. Pemberton.

Bourguet, Louis. 1729. *Lettres philosophiques sur la formation des sels et des crystaux et sur la génération et le mechanisme organique des plantes et des animaux.* Amsterdam: F. l'Honoré.

————. 1742a. *Mémoires pour servir à l'histoire naturelle des petrifications dans les quatre parties du monde.* La Haye: J. Neaulme.

————. 1742b. *Traité des petrifications.* Paris: Briasson.

Bowen, N. L. 1928. *The evolution of the igneous rocks.* Princeton, N. J.: Princeton Univ. Press.

Bowler, Peter. 1976. *Fossils and progress.* New York: Science History Publications.

Boyle, Robert. 1669. *Certain physiological essays and other tracts.* 2d ed. (1st ed., 1661). Oxford: R. Davis.

————. [1672] 1972. *An essay about the origine and virtues of gems.* Reprint. New York: Hafner.

Bridson, Gavin, and Anthony P. Harvey. 1971. A checklist of natural history bibliographies and bibliographical scholarship, 1966–1970. *Journal of the Society for the Bibliography of Natural History.* 5:428–67.

Brochant de Villiers, A.-J.-F.-M. 1801–2. *Traité élémentaire de minéralogie, suivant les principes du professeur Werner.* 2 vols. Paris.

Brocchi, Giovanni. 1814. *Conchiologia fossile subapennina con osservazioni geologiche sugli Apennini e sul suolo adiacente.* Milan: Royal Printing Office.

Brongniart, Alexandre. 1807. *Traité élémentaire de minéralogie.* 2 vols. Paris.

————. 1816–30. "Volcans." In *Dictionnaire des sciences naturelles,* 58:437–43. Strasbourg: Levrault.

————. 1821. Sur les caractères zoologiques des formations, avec l'application de ces caractères à la détermination de quelques terrains de craie. *Annales des Mines.* 16:537–72.

————. 1827. *Classification et caractères minéralogiques des roches homogènes et hétérogènes.* Paris and Strasbourg: F. G. Levrault.

Brooke, John. 1979. "The natural theology of the geologists: Some theological strata." In *Images of the earth: Essays in the history of the environmental sciences,* ed. L. Jordanova and R. S. Porter, 39–66. Chalfont St. Giles: British Society for the History of Science.

Buch, Leopold von. 1808. Über das Fortschritten der Bildungen in der Natur. *Moll's Ephemeriden der Berg- und Hüttenkunde.* 4:1–16.

————. 1816. Von den geognostischen Verhältnissen des Trapp-Porphyry. *Abhandlungen der Königlichen Akademie der Wissenschaften, Berlin.* 129–51.

————. 1820. Über die Zusammensetzung der basaltischen Inseln und über Erhebungs-Cratere. *Abhandlungen der Königlichen Akademie der Wissenschaften, Berlin.* 51–86.

————. 1824. Über die geognostichen System von Deutschland." *Leonhard's Mineralogisches Taschenbuch,* 501–6. Frankfurt.

————. 1825. Über Dolomit als Gebirgsart. *Abhandlungen der Königlichen Akademie der Wissenschaften, Berlin.* 83–136.

————. 1836. Über Erhebungskratere und Vulkane. *Poggendorf's Annalen der Physik und Chemie.* 37:169–90.

————. 1867. *Gesammelte Schriften.* 4 vols. Berlin: Reimer.

Buckland, William. 1821. Notice of a paper laid before the Geological Society on the structure of the Alps and adjoining parts of the continent, and their relation to the secondary and transition rocks in England. *Annals of Philosophy.* 1:450–68.

————. 1823. *Reliquiae Diluvianae.* London.

————. 1829. On the formation of the valley of Kingsclere and other valleys by the elevation of the strata that inclose them. *Transactions of the Geological Society of London.* 2, new ser.:119–30.

Buffon, Georges-Louis Leclerc, Comte de. 1884–85. *Oeuvres completes.* Edited by J. L. Lanessan. Paris.

Burchfield, Joe D. 1975. *Lord Kelvin and the age of the earth.* New York: Science History.

Burke, John G. 1966. *Origins of the science of crystals.* Berkeley: Univ. of California Press.

————. 1969. Mineral classification in the early nineteenth century. In *Toward a History of Geology,* ed. C. Schneer, 62–77. Cambridge, Mass.: MIT Press.

Burnet, Thomas. 1681/1961. *The sacred theory of the earth.* Carbondale: Southern Illinois Univ. Press.

Burt, R. 1969. *Cornish mining.* Newton Abbott: David and Charles.

Butterfield, Herbert. 1955. *Man on his past: The study of the history of historical scholarship.* Cambridge: Cambridge Univ. Press.

Butts, Robert E., and Davis, John W. 1970. *The methodological heritage of Newton.* Toronto: Univ. of Toronto Press.

Cain, A. J. 1958. Logic and memory of Linnaeus's system of taxonomy. *Proceedings of the Linnaean Society of London.* 169:144–63.

Cannon, Susan F. 1978. *Science in culture.* New York: Science History.

Cannon, Walter. 1960. The uniformitarian-catastrophist debate. *Isis.* 51:38–55.

————. 1961. The impact of uniformitarianism: Two letters from John Herschel to Charles Lyell, 1836–1837. *Proceedings of the American Philosophical Society.* 105:301–14.

————. 1964. History in depth: The early Victorian period. *History of Science.* 3:20–38.

————. 1976. Charles Lyell, radical actualism and theory. *British Journal for the History of Science.* 9:104–20.

Cardwell, D. L. 1972. rev. ed. *The organisation of science in England.* London: Heinemann.

Carozzi, Albert V. 1960. A study of Werner's personal copy of *Von den äusserlichen Kennzeichen der Fossilien*, 1774. *Isis.* 51:554–57.

———. 1965. Lavoisier's fundamental contribution to stratigraphy. *Ohio Journal of Science.* 65:65–84.

———. 1969a. The geological contribution of Rudolf Erich Raspe (1737–1794). *Archives des Sciences, Genève.* 22:625–43.

———. 1969b. Rudolf Erich Raspe and the basalt conroversy. *Studies in Romanticism.* 8:235–50.

———. 1970. Robert Hooke, Rudolf Erich Raspe, and the concept of "earthquakes." *Isis.* 61:85–91.

Carozzi, Albert V., and Donald H. Zenger. [1791] 1981. Trans. of "Sur un genre de pierres calcaires tres-peu effervescentes avec les acides," *Journal of Geological Education.* 29:4–10.

Carozzi, Marguerite. 1983. "Voltaire's attitude to geology." *Archives des Sciences, Genève.* 36:1–145.

Carozzi, Marguerite, and Carozzi, A. V. 1984. Élie Bertrand's changing theory of the earth. *Archives des Sciences, Genève.* 37:265–300.

Cesalpino, Andrea. 1596. *De metallicis libri tres.* Rome: A. Zannetti.

Challinor, John. 1949. Thomas Webster's observations on the geology of the Isle of Wight, 1811–13. *Proceedings of the Isle of Wight Natural History and Archaeological Society.* 4:108–22.

———. 1961–64. Some correspondence of Thomas Webster, geologist (1773–1844). *Annals of Science.* 17:175–95; 18:147–75; 19:49–79, 285–97; 20:59–80, 143–64.

Charpentier, Johann Friedrich Wilhelm von. 1778. *Mineralogische Geographie der Chursächsischen Lande.* Leipzig: S. L. Crusius.

———. 1799. *Beobachtungen über die Lagerstätte der Erze hauptsächlich aus den sächsischen Gebirgen.* Leipzig: G. J. Göschen.

Charpentier, Jean de. 1823. *Essai sur la constitution géognostique des Pyrénées.* Paris: F.-G. Levrault.

Chevalier, Jean. 1947. La mission de Gabriel Jars dans les mines Brittaniques en 1764. *Transactions of the Newcomen Society.* 57–68.

Chitnis, A. C. 1970. The University of Edinburgh's natural history museum and the Huttonian-Wernerian debate. *Annals of Science.* 26:85–94.

Christie, J. R. 1981. Ether and the science of chemistry: 1740–1790. In *Conceptions of ether*, ed. G. N. Cantor and M. J. S. Hodge. Cambridge: Cambridge Univ. Press.

Cleaveland, Parker. 1816. *An elementary treatise on mineralogy and geology.* 2d ed. 2 vols. Boston.

Cohen, I. Bernard. 1956. *Franklin and Newton.* Philadelphia: American Philosophical Society.

Coleman, William. 1963. Abraham Gottlob Werner vu par Alexander von Humboldt, avec des notes de Georges Cuvier. *Sudhoffs Archiv für Geschichte der Medizin und der Naturwissenschaften.* 47:465–78.

————. 1964. *Georges Cuvier, zoologist*. Cambridge, Mass.: Harvard Univ. Press.

————. 1973. Limits of the recapitulation theory: Carl Friedrich Kielmeyr's critique of the presumed parallelism of earth history, ontogeny, and the present order of organisms. *Isis*. 64:341–50.

Collier, Katherine B. 1934. *Cosmogonies of our fathers: Some theories of the seventeenth and the eighteenth centuries*. New York: Columbia University Press.

Conybeare, W. D. 1816. Descriptive notes referring to the outline sections presented by a part of the coasts of Antrim and Derry. *Transactions of the Geological Society of London*. 2:196–216.

————. 1823. Memoir illustrative of a general geological map of the principal mountain chains of Europe. *Annals of Philosophy*. new ser. 5:1–16, 135–49, 210–18, 278–89, 356–59; 6: 214–19.

————. 1829. Answer to Dr. Fleming's view of the evidence from the animal kingdom, as to the former temperature of the northern regions. *Edinburgh New Philosophical Journal*. 7: 142–52.

————. 1833. Report on the progress, actual state and ulterior prospects of geological science. *Report of the British Association for the Advancement of Science for 1831–2*. 365–414.

Conybeare, W. D., and Phillips, W. 1822. *Outlines of the geology of England and Wales*. London: Phillips.

Cordier, Pierre-Louis-Antoine. 1816. Sur les substances minérales dites en masse dans la composition des roches volcaniques. *Journal de Physique*. 82:261–68; 83:185–261, 284–307, 352–86.

————. 1827. Essai sur la température de l'intérieur de la terre. *Mémoires de l'Académie des Sciences, Paris*. 7:473–556.

Cox, L. R. 1942. New light on William Smith and his work. *Proceedings of the Yorkshire Geological Society*. 25:1–99.

Craig, G. Y., ed. 1978. *James Hutton's theory of the earth: The lost drawings*. Edinburgh: Scottish Academic Press.

Cronstedt, Axel Fredrik. [1758] 1770. *An essay towards a system of mineralogy*. Trans. from German edition. London: E. and C. Dilly.

Crosland, Maurice P. 1962. *Historical studies in the language of chemistry*. Cambridge, Mass.: Harvard Univ. Press.

Cuvier, Georges. 1799. Mémoire sur les espèces d'éléphans vivantes et fossiles. *Mémoires de l'Institut national des sciences et arts, sciences mathématiques et physiques*. 2:1–22.

————. 1810. *Rapport historique sur les progrès des sciences naturelles depuis 1789*. Paris: Imprimerie Imperiale.

————. 1812. *Recherches sur les ossemens fossiles de quadrupèdes*. Paris: Deterville.

————. [1817] 1978. *Essay on the theory of the earth: With mineralogical notes, and an account of Cuvier's geological discoveries by Professor Jameson*. Reprint of 3d ed. (1st ed., 1813). New York: Arno.

Cuvier, Georges, and Brongniart, A. 1808. Essai sur la géographie minéralogique des environs de Paris. *Journal des Mines.* 23:421–58.

———. 1811. *Essai sur la géographie minéralogique des environs de Paris avec une carte géognostique, et des coupes de terrain.* Paris: Bandouin.

Dance, Peter. 1966. *Shell collecting: An illustrated history.* London: Faber and Faber.

Darcet, Jean. 1766. *Mémoire sur l'action d'un feu égal, violent, et continué pendant plusieurs jours sur un grand nombre de terres, de pierres & de chaux métalliques.* Paris: P. G. Cavelier.

Darwin, Charles. 1837a. Observations of proofs of recent elevation on the coast of Chile. *Proceedings of the Geological Society of London.* 2:446–49.

———. 1837b. On certain areas of elevation and subsidence in the Pacific and Indian Oceans, as deduced from the study of coral formations. *Proceedings of the Geological Society of London.* 2:552–54.

———. 1840. On the connexion of certain volcanic phenomena in South America; and on the formation of mountain chains and volcanos, as the effect of the same power by which continents are elevated. *Transactions of the Geological Society of London.* 2d ser. 5:601–31.

———. [1844] 1897. *Geological observations on the volcanic islands and parts of South America visited during the voyage of H.M.S. Beagle.* 3d ed. New York: Appleton.

Daubeny, Charles. 1826. *A description of active and extinct volcanos.* London: Phillips.

Davies, Gordon L. 1969. *The earth in decay: A history of British geomorphology, 1578–1878.* New York: Macdonald.

Davy, Humphry. 1828. On the phenomena of volcanoes. *Philosophical Transactions of the Royal Society.* 118:241–50.

Dean, Dennis R. 1969. Edward Jorden and the fermentation of the metals: An iatrochemical study of terrestrial phenomena. In *Toward a history of geology,* ed. C. Schneer, 100–21. Cambridge, Mass.: MIT Press.

———. 1973. James Hutton and his public, 1785–1805. *Annals of Science.* 30:89–105.

———. 1979. The word "geology." *Annals of Science.* 36:35–43.

———. 1980. Graham Island, Charles Lyell, and the craters of elevation controversy. *Isis.* 71:571–88.

———. 1981. The age of the earth controversy: Beginnings to Hutton. *Annals of Science.* 38:435–56.

Debus, Allen. 1977. *The chemical philosophy: Paracelsian science and medicine in the sixteenth and seventeenth centuries.* New York: Science History.

Dechen, H. von. 1853. *Leopold von Buch, sein Einfluss auf die Entwicklung der Geognosie.* Bonn.

De la Beche, Henry, ed. 1824. *A selection of geological memoirs contained in the Annales des Mines, written by Brongniart, Humboldt, Von Buch and others.* London: Phillips.

———. 1829. Note on the differences, either original or consequent on disturbance, which are observable in the secondary stratified rocks. *Philosophical Magazine.* 6:213–25.

———. 1830. *Sections and views, illustrative of geological phaenomena.* London.

———. 1834. On the geology of Nice. *Proceedings of the Geological Society of London.* 1:87–89.

———. 1851. *The geological observer.* London: Longman.

del Rio, A. M. 1795–1805. *Elementos de Orictognosia.* 2 vols. Mexico City.

Deluc, J. A. 1779–80. *Lettres physiques et morales sur l'histoire de la terre et de l'homme.* 5 vols. Paris.

———. 1790. Letters to Dr. James Hutton on his theory of the earth. *Monthly Review.* 2:206–27,582–95; 3:573–86; 5:564–85.

———. 1809. *An elementary treatise on geology.* London: Rivington.

Descartes, Réné. [1644] 1984. *Principles of philosophy.* Trans. Valentine R. Miller and Reese P. Miller. Dordrecht, Netherlands: Reidel.

Deshayes, Gerald Paul. 1824–37. *Description des coquilles fossiles des environs de Paris.* 2 vols. and atlas. Paris.

Desmarest, N. 1774. Mémoire sur l'origine et la nature du basalte. *Mémoires de l'Académie Royale des Sciences, Paris.* 705–75.

———. 1794/5–1828). *Géographie physique.* 5 vols. [Encyclopèdie Méthodique]. Paris: H. Agasse.

———. 1806. "Mémoire sur la détermination des trois époques de la nature," *Mémoires de l'Institut, Paris.* 6:219–89.

Dolomieu, Déodat Guy Silvain de. 1790. Lettre . . . sur la question de l'origine du basalte. *Observations sur la physique.* 37:193–202.

———. 1791. Mémoire sur les pierres composés et sur les roches . . . *Observations sur la physique.* 39:374–407.

———. 1801. *Sur la philosophie minéralogique et sur l'espèce minéralogique.* Paris: Bossange, Masson et Besson.

Donald, M. B. 1961. *Elizabethan monopolies: The history of the Company of Mineral and Battery Works from 1565–1604.* Edinburgh: Oliver and Boyd.

Donovan, Arthur. 1975. *Philosophical chemistry in the Scottish enlightenment.* Edinburgh: Edinburgh Univ. Press.

———. 1977. James Hutton and the Scottish Enlightenment: Some preliminary considerations. *Scotia.* 1:56–68.

———. 1978. James Hutton, Joseph Black, and the chemical theory of heat. *Ambix.* 25:176–90.

Donovan, Arthur, and Prentiss, Joseph. 1980. James Hutton's medical dissertation. *Transactions of the American Philosophical Society.* 17(6):1–57.

Dott, Robert H. 1969. James Hutton and the concept of a dynamic earth. In

Toward a history of geology, ed. C. J. Schneer, 122–41. Cambridge, Mass.: MIT Press.

Duckham, B. F. 1970. *A history of the Scottish coal industry*. Newton Abbott: David and Charles.

Duhem, Pierre. [1906] 1954. *The aim and structure of physical theory*. Trans. P. P. Wiener. Princeton, N.J.: Princeton Univ. Press.

———. 1913–59. *Le système du monde: Histoire des doctrines cosmologiques de Platon à Copernic*. 10 vols. Paris.

Durkheim, Emile, and M. Mauss. [1903] 1963. *Primitive classification*. Trans. R. Needham from "De quelques forms primitives de classification," *Année Sociologique*. London: Cohen and West.

Eichholz, D. E. 1949. Aristotle's theory of the formation of metals and minerals. *Classical Quarterly*. 43:141–46.

Élie de Beaumont, L. 1829. *Recherches sur quelques-unes des révolutions de la surface du globe*. Paris: Crochard.

———. 1831. Researches on some of the revolutions which have taken place on the surface of the globe. . . . *Philosophical Magazine*. new ser. 16:241–64 (Translation of a 1829 article.)

Ellenberger, François. 1984. Louis Cordier (1777–1861), initiateur de l'étude microscopique des laves: Percées sans lendemain ou innovation decisive. *Earth Sciences History*. 3:44–53.

Emerton, Norma. 1984. *The scientific reinterpretation of form*. Ithaca, N. Y.: Cornell Univ. Press.

Englefield, Henry. 1816. *A description of the principal picturesque beauties, antiquities, and geological phenomena of the Isle of Wight*. London.

Engelhardt, W. von. 1980. Carl von Linné und das Reich der Steine. In *Carl von Linné: Beitrag über Zeitgeist, Werk und Wirkungsgeschichte*, ed. H. Goerke et al. Göttingen: Vandenhoeck & Ruprecht.

Eyles, Joan. 1969. William Smith: Some aspects of his life and work. In *Toward a history of geology*, ed. C. Schneer, 142–58. Cambridge, Mass.: MIT Press.

Eyles, Victor A. 1950. Note on the original publication of Hutton's *Theory of the Earth* and on the subsequent forms in which it was issued. *Proceedings of the Royal Society of Edinburgh*. 13B:377–86.

———. 1955. A bibliographical note on the earliest printed version of James Hutton's *Theory of the Earth*, its form and date of publication. *Journal of the Society for Bibliography of Natural History*. 3:105–8.

———. 1958. The influence of Nicholas Steno on the development of geological science in Britain. In *Nicholas Steno and his Indice*, ed. G. Scherz, 167–88. Copenhagen.

———. 1961. Sir James Hall, Bt. (1761–1832). *Endeavour*. 20:216.

———. 1964. Abraham Gottlob Werner (1749–1817) and his position in the history of the mineralogical and geological sciences. *History of Science*. 3:102–15.

———. 1966. The history of geology. *History of Science*. 5:77–86.

———. 1969. The extent of geological knowledge in the eighteenth century, and the methods by which it was diffused. In *Toward a history of geology*, ed. C. Schneer, 159–83. Cambridge, Mass.: MIT Press.

———. 1972. Mineralogical maps as forerunners of modern geological maps. *Cartographic Journal*. 9:133–35.

Eyles, Victor A., and Joan M. Eyles. 1951. Some geological correspondence of James Hutton. *Annals of Science*. 7:323–25.

Farber, Paul. 1972. Buffon and the concept of species. *Journal of the History of Biology*. 5:259–84.

Farey, John. 1811. *General view of the agriculture and minerals of Derbyshire*. London.

———. 1815. Observations on the priority of Mr. Smith's investigations of the strata of England; on the very unhandsome conduct of certain persons in detracting from his merit therein; and the endeavour of others to supplant him in the sale of his maps. *Philosophical Magazine*. 45:333–44.

Faujas de Saint-Fond, B. 1778. *Recherches sur les volcans éteints du Vivarais et du Velay*. Grenoble.

———. 1788. *Essai sur l'histoire naturelle des roches de trapp, contenant leur analyse, et des recherches sur leurs caractères distinctifs*. Paris.

Ferguson, Adam. 1805. Minutes of the life and character of Joseph Black, M.D. *Transactions of the Royal Society of Edinburgh*. 5:101–17.

Fitton, William. 1802. Review of Playfair's *Illustrations of the Huttonian theory of the earth*. *Edinburgh Review*. 1:201–16.

———. 1818. Transactions of the Geological Society of London, vol. III. *Edinburgh Review*. 29:70–94.

———. 1832–33. Notes on the history of English geology. *Philosophical Magazine*. 1:147–60, 268–75, 442–50; 2:35–57.

Fleming, John. 1829. On the value of the evidence from the animal kingdom, tending to prove that the arctic regions formerly enjoyed a milder climate than at present. *Edinburgh New Philosophical Journal*. 6:277–86.

———. 1829. Additional remarks on the climate of the arctic regions, in answer to Mr. Conybeare. *Edinburgh New Philosophical Journal*. 8:65–74.

Foucault, Michel. [1966] 1973. *The order of things: An archaeology of the human sciences*. Trans. from *Les mots et les choses*. New York: Vintage Books.

Fourier, J. B. L. 1819. Mémoire sur le refroidissement séculaire du globe terrestre. *Annales de Chimie*. 13:418–37.

———. 1827. Mémoire sur les températures du globe terrestre et des éspaces planetaire. *Mémoires de l'Académie des Sciences, Paris*. 569–604.

Frängsmayr, Tore. 1969. *Geologi och skapelsetro: Föreställningar om jordens historia från Hiärne till Bergman*. English summary. Stockholm: Almqvist and Wiksell.

———. 1974. Swedish science in the eighteenth century. *History of Science*. 12:29–42.

————. 1976. The geological ideas of J. J. Berzelius. *British Journal for the History of Science.* 9:228–36.

————, ed. 1983a. *Linnaeus: The man and his work.* Berkeley: Univ. of California Press.

————. 1983b. Linnaeus as geologist. In *Linnaeus: The man and his work,* ed. Tore Frängsmayr. Berkeley: Univ. of California Press.

Freiesleben, Johann Karl. 1807–15. *Geognostischer Beitrag zur Kenntnis des Kupferschiefergebirges.* 4 vols. Freiberg.

Freyberg, Bruno von. 1955. Johann Gottlob Lehmann (1719–1767): Ein Arzt, Chemiker, Metallurg, Bergmann, Mineraloge und grundlegender Geologe. *Erlanger Forschungen: Reihe B. Naturwissenschaften.* 1.

Galbraith, Winslow H. 1974. James Hutton: An analytic and historical study. Ph.D. Diss., Univ. of Pittsburgh.

Gay-Lussac, Joseph-Louis. 1823. Reflexions sur les volcans. *Annales de Chimie.* 22:415–29.

Geikie, Archibald. [1905] 1962. *The founders of geology.* Reprint. New York: Dover.

Gellert, Christlieb Ehregott. 1776. *Metallurgic chymistry: Being a system of mineralogy in general, and of all the arts arising from this science.* London: T. Becket.

Geological inquiries. [1808] 1817. Reprinted in the *Philosophical Magazine.* 40:420–29.

Gerstner, Patsy. 1968. James Hutton's theory of the earth and his theory of matter. *Isis.* 59:26–31.

————. 1971. The reaction to James Hutton's use of heat as a geological agent. *British Journal for History of Science.* 5:353–62.

Gesner, Konrad. 1565. *De rerum fossilium.* Tiguri.

Gillispie, C. C. [1951] 1959. *Genesis and geology.* Reprint. New York: Harper.

————. 1980. *Science and polity in France at the end of the old regime.* Princeton, N.J.: Princeton Univ. Press.

Gohau, Gabriel. 1983. "Idées anciennes sur la formation des montagnes." *Cahiers d'historie et de philosophie des science.* 7:1–86.

Goodman, D. C. 1971. The application of chemical criteria to biological classification in the eighteenth century. *Medical History.* 15:23–44.

Gortani, M. 1963. Italian pioneers in geology and mineralogy. *Journal of World History.* 7:503–19.

Gough, J. W. [1967] 1930. *The mines of Mendip.* Reprint. New York: Augustus Kelley.

Gould, Stephen J. 1981. This view of life. *Natural History.* 90:20–24.

Grant, R. 1979. Hutton's theory of the earth. In *Images of the Earth,* ed. Ludmilla Jordanova and Roy Porter, 23–38. Chalfont St. Giles Bucks.: British Society for the History of Science.

Greene, John. [1959] 1961. *The death of Adam: Evolution and its impact on western thought.* Reprint. New York; Mentor.

————. 1967. Review of M. Foucault's *Les Mots et les Choses. Social Science Information.* 6(4):131–38.

————. 1969. The development of mineralogy in Philadelphia, 1780–1820: A summary. In *Toward a history of geology*, ed. C. J. Schneer, 184–85. Cambridge, Mass.: MIT Press.

Greene, John, and John G. Burke. 1978. The science of minerals in the age of Jefferson. *Transactions of the American Philosophical Society.* 68: 1–113.

Greene, Mott. 1983. *Geology in the nineteenth century.* Ithaca, N.Y.: Cornell Univ. Press.

Greenough, George Bellas. 1819. *A critical examination of the first principles of geology.* London: Longman.

————. 1834. Remarks on the theory of the elevation of mountains. *Edinburgh New Philosophical Journal.* 17:205–27.

Guettard, Jean-Étienne. 1751. Mémoire et carte minéralogique sur la nature et la situation des terreins qui traversent la France et l'Angleterre. *Mémoires de l'Academie Royale des Sciences, 1746.* 363–92.

————. 1756. Mémoire sur quelques montagnes de la France qui ont été des volcans. *Mémoires de l'Académie Royale des Sciences, 1752.* 27–59.

————. 1768–86. *Mémoires sur differentes parties de la physique, de l'histoire naturelle, des sciences et des arts.* 5 vols. Paris.

Guettard, Jean-Étienne, and A. Monnet. 1780. *Atlas et description minéralogiques de la France.* Paris.

Gunnar, Eriksson. 1983. Linnaeus the botanist. In *Linnaeus: The man and his work*, ed. Tore Frängsmayr. Berkeley: Univ. of California Press.

Guntau, Martin. 1967. Der Aktualismus bei A. G. Werner. *Bergakademie.* 5:296–97.

————. 1969. Die Entwicklung der Vorstellungen von der Mineralogie in der Wissenschaftsgeschichte. *Geologie.* 18:526–37.

————. 1971. Kriterien für die Herausbildung der Lagerstättenlehre als Wissenschaft im 19. Jahrhundert. *Geologie.* 20:348–61.

————.1974a. Die geologische Wissenschaften an der Bergakademie Freiberg in der Periode der industriellen Revolution. *NTM-Zeitschrift für Geschichte der Naturwissenschaft, Technik und Medizin.* 11:16–23.

————. 1974b. Leopold von Buch: Gedanken zu seinen Leben und Wirken als Geologie. *Zeitschrift für geologische Wissenschaft.* 2:1371–83.

————. 1974c. Leopold von Buch 1774–1853: Kolloquium aus Anlass der Wiederkehr seines Geburtstages nach 200 Jahren. *Zeitschrift für geologische Wissenschaft.* 2:1363–65.

————. 1981–82. Les rélations des géologues et minéralogistes de l'école de Freiberg avec les érudits français. *Histoire et Nature.* 19–20:107–14.

————. 1984. *Die Genesis der Geologie als Wissenschaft: Studie zu den kognitiven Prozessen und gesellschaftlichen Bedingungen bei der Herausbildung der Geologie als naturwissenschaftliche Disziplin an der Wende vom 18. zum 19. Jahrhundert.* Berlin: Akademie Verlag.

Guntau, Martin, and Wolfgang Mühlfriedel. 1968. Die Bedeutung von Abraham Gottlob Werner für die Mineralogie und die Geologie. *Geologie.* 17:1096–1115.

Haber, Francis. 1959. *The age of the world: Moses to Darwin.* Baltimore: Johns Hopkins Univ. Press.

Haidinger, Karl. 1787. *Systematische Eintheilung der Gebirgsarten.* Vienna: Wappler.

Hales, Stephen. 1750. *Some considerations on the causes of earthquakes.* London: R. Manby and H. S. Cox.

Hall, A. Rupert. 1983. On Whiggism. *History of Science.* 21:45–59.

Hall, James. 1794. Sir James Hall on granite. *Transactions of the Royal Society of Edinburgh.* 3:8–12.

———. 1798. Curious circumstances upon which the vitreous or the stony character of whinstone and lava respectively depend. *[Nicholson's] Journal of Natural Philosophy.* 2:285–88.

———. 1804. Experiments on the effects of heat modified by compression. *[Nicholson's] Journal of Natural Philosophy.* 9:98–107.

———. 1805. Experiments on whinstone and lava. *Transactions of the Royal Society of Edinburgh.* 5:43–75.

———. 1812. Account of a series of experiments, showing the effect of compression in modifying the effects of heat. *Transactions of the Royal Society of Edinburgh.* 6:71–185.

———. 1815a. On the revolutions of the earth's surface. *Transactions of the Royal Society of Edinburgh.* 7:139–212.

———. 1815b. On the vertical position and convolutions of certain strata, and their relation with granite. *Transactions of the Royal Society of Edinburgh.* 7:79–108.

———. 1826. On the consolidation of the strata of the earth. *Transactions of the Royal Society of Edinburgh.* 10:314–29.

Hall, Marie Boas. 1965. *Robert Boyle on natural philosophy.* Bloomington: Indiana Univ. Press.

Hallam, Anthony. 1978. *After the revolution.* Birmingham: Univ. of Birmingham Press.

———. 1983. *Great geological controversies.* Oxford: Oxford Univ. Press.

Hamilton, William. 1772. *Observations on Mount Vesuvius, Mount Etna, and other volcanoes.* London: T. Cadell.

Harris, John. 1704. *Lexicon technicum.* London

Haüy, René Just. 1784. *Essai d'une théorie sur la structure des crystaux, appliquée à plusieurs genres de substances cristallisées.* Paris: Gogue & Nee de la Rochelle.

———. 1801. *Traité de minéralogie.* Paris: Louis.

———. 1807. *An elementary treatise on natural philosophy.* London: G. Kearsley.

———. 1809. *Tableau comparatif des résultats de la cristallographie et de l'analyse chimique, relativement à la classification des minéraux.* Paris: Courcier.

———. 1822. *Traité de cristallographie, suivi d'une application des principes de cette science à la détermination des espèces minérales.* 2 vols. and atlas. Paris: Bachelier et Huzard.

Hedberg, Hollis D. 1969. The influence of Torbern Bergman (1735–1784) on stratigraphy: A resumé. In *Toward a history of geology,* ed. C. J. Schneer, 186–192. Cambridge, Mass.: MIT Press.

Heim, Johann Ludwig. 1796–1812. *Geologische beschreibung des Thüringer Waldgeburgs.* 6 vols. Meiningen.

Heimann, P. M. 1981. Ether and imponderables. In *Conceptions of ether,* ed. G. N. Cantor and M. J. S. Hodge. Cambridge: Cambridge Univ. Press.

Heimann, P. M., and J. E. McGuire. 1971. Newtonian forces and Lockean powers: Concepts of matter in eighteenth-century thought. *Historical Studies in the Physical Sciences.* 3:233–306.

Henckel, Johann Friedrich. 1722. *Flora saturnizans.* Leipzig.

———. 1725. *Pyritologia.* Leipzig.

———. 1794–95. *Mineralogische, chemische und alchymistische Briefe.* 3 vols. Dresden.

Henderson, William Otto. 1958. *The state and the industrial revolution in Prussia, 1748–1870.* Liverpool: Liverpool Univ. Press.

———. 1968. *The industrial revolution in Europe.* Chicago: Quadrangle Books.

Herbert, Sandra. 1970. The logic of Darwin's discovery. Ph.D. diss., Brandeis Univ.

———, ed. 1980. *The red notebook of Charles Darwin.* Ithaca, N.Y., and London: Cornell Univ. Press.

Herrmann, W. 1953. Die Enstehung der Freiberger Bergakademie. *Freiberger Forschungshefte: Kultur und Technik.* D2:23–42.

———. 1962. Bergrat Henckel: Ein Wegbereiter der Bergakademie. *Freiberger Forschungshefte: Kultur und Technik.* D37.

Herschel, John. 1825. Notice of a remarkable occurrence of serpentine at the junction of sienite with the dolomite of Tyrol. *Edinburgh Journal of Science.* 3:126–29.

———. [1831] 1961. *Preliminary discourse on the study of natural philosophy.* Reprint. New York: Johnson.

———. 1832. On the astronomical causes which may influence geological phenomena. *Transactions of the Geological Society of London.* 3:293–99.

Hesse, Mary. 1974. *The structure of scientific inference.* Berkeley: Univ. of California Press.

Hill, John. 1748. *A general natural history; or, new and accurate descriptions of the animals, vegetables, and minerals, of the different parts of the world.* London: T. Osborne.

Hodge, M. J. S. 1983. Darwin and the laws of the animate part of the terrestrial system (1835–1837): On the Lyellian origins of his zoonomical explanatory program. *Studies in History of Biology.* 6:1–106.

Hoff, K. E. A. von. 1810. Beobachtungen über die Verhältnisse des

Basaltes an einigen Bergen von Hessen und Thüringen. *Magazin der Gesellschaft naturforschenden Freunde zu Berlin.* 5:347–62.

———. 1822–41. *Geschichte der durch Überlieferung nachgewiesenen natürlichen Veränderungen der Erdoberfläche.* 5 vols. Gotha.

Hoffmann, Friedrich. 1832. *Über die geognistische Beschaffenheit der Liparischen Inseln.* Leipzig.

———. 1838. *Geschichte der Geognosie, und Schilderung der vulkanischen Erscheinungen.* Berlin.

Hölder, H. 1960. *Geologie und Paläontologie in Texten und Geschichte.* Freiburg.

Holmes, F. L. 1971. Analysis by fire and solvent extractions: the metamorphosis of a tradition. *Isis.* 62:129–48.

Holmyard, E. J., and D. C. Mandeville. 1927. *Avicennae de congelatione et conglutinatione lapidum.* Paris: Libraire Orientaliste.

Hooke, Robert. [1665] 1961. *Micrographia or some physiological descriptions of minute bodies made by magnifying glasses and observations and inquiries thereupon.* Reprint. New York: Dover.

———. (1705), 1971. *The posthumous works of Robert Hooke.* Ed. S. Smith and B. Walford. London: Frank Cass.

Hookyaas, R. 1947. The discrimination between "natural" and "artificial" substances and the development of the corpuscular theory. *Archives International d'Histoire des Sciences.* 1:640–51.

———. 1952a. The species concept in eighteenth-century mineralogy. *Archives International d'Histoire des Sciences.* 5:45–55.

———. 1952b. Torbern Bergman's crystal theory. *Lychnos.* 21–54.

———. 1953. *La naissance de la cristallographie en France au XVIIIe siècle.* Paris: Université de Paris.

———. 1955. Les debuts de la théorie cristallographique de R. J. Haüy, d'après les documents originaux. *Revue d'histoire des sciences.* 8:319–37.

———. 1959. *Natural law and divine miracle: A historical-critical study of the principle of uniformity in geology, biology, and theology.* Leiden: E. J. Brill.

Hopkins, William. 1838. Researches in physical geology. *Transactions of the Cambridge Philosophical Society.* 6:1–84.

Hufbauer, Karl. 1982. *The formation of the German chemical community, 1720–1795.* Berkeley: Univ. of California Press.

Hull, David. 1972. Charles Darwin and nineteenth-century philosophies of science. In *Foundations of scientific method: The nineteenth century,* ed. Ronald Giere and Richard Westfall, 115–32. Bloomington: Indiana Univ. Press.

———. 1973. *Darwin and his critics: The reception of Darwin's theory of evolution by the scientific community.* Cambridge, Mass.: Harvard Univ. Press.

———. 1974. *The philosophy of biological science.* Englewood Cliffs, N.J.: Prentice-Hall.

———. 1976. Are species individuals? *Systematic Zoology.* 25:174–91.

──────. 1978. A matter of individuality, *Philosophy of Science*. 45:33–60.

──────. 1979. In defense of presentism. *History and Theory*. 18:1–15.

Humboldt, Alexander von. 1790. *Mineralogische Beobachtungen über einige Basalte am Rhein*. Brunswick.

──────. 1810. Essai sur la Nouvelle Espagne. Trans. in the *Journal of Natural Philosophy, Chemistry and the Arts*. 26:81–86.

──────. 1820–21. On isothermal lines, and the distribution of heat over the globe. *Edinburgh Philosophical Journal*. 3:1–20, 256–274; 4:23–38, 262–281; 5:28–39.

──────. 1823. *A geognostical essay on the superposition of rocks in both hemispheres*. Trans. of *Essai géognostique sur le gisement des roches dans les deux hémisphères*. London: Longman.

Humboldt, Alexander von. 1845–62. *Kosmos*. 5 vols and atlas. Stuttgart: Cotta.

Hutton, James. 1788. Theory of the earth; or an investigation of the laws observable in the composition, dissolution, and restoration of land upon the globe. *Transactions of the Royal Society of Edinburgh*. 1:209–304.

──────. 1792. *Dissertations on different subjects in natural philosophy*. Edinburgh: Strahan and Cadell.

──────. 1794a. Observations on granite. *Transactions of the Royal Society of Edinburgh*. 3:77–81.

──────. 1794b. *An investigation of the principles of knowledge, and of the progress of reason, from sense to science and philosophy*. Edinburgh: Strahan and Cadell.

──────. 1794c. *A dissertation upon the philosophy of light, heat, and fire*. Edinburgh: Cadell and Davies.

──────. 1795. *Theory of the earth with proofs and illustrations*. Edinburgh: Creech.

Jacob, Margaret C. 1976. *The Newtonians and the English Revolution, 1689–1720*. Ithaca, N.Y.: Cornell Univ. Press.

Jacob, Margaret C., and W. A. Lockwood. 1972. Political millenarianism and Burnet's *Sacred Theory*. *Science Studies*. 2:265–79.

Jameson, Robert. [1804–8] *System of mineralogy: comprehending oryctognosie, geognosie, mineralogical chemistry, mineralogical geography, and economical mineralogy*. Edinburgh: A. Constable.

──────. [1808] 1976. *The Wernerian theory of the neptunian origin of rocks: Elements of geognosy, 1808*. Facsimile reprint of *System of mineralogy*, vol. 3., with an introduction by J. M. Sweet. New York: Hafner.

──────. 1811. On colouring geognostical maps. *Memoirs of the Wernerian Natural History Society*. 1:149–61.

Jars, Gabriel, ed. 1774–81. *Voyages métallurgiques, ou recherches et observations sur les mines et forges de fer*. 3 vols. Lyons and Paris.

Jordanova, Ludmilla, and Roy Porter, eds. 1979. *Images of the earth: Essays in the history of the environmental sciences*. Chalfont St. Giles: British Society for the History of Science.

Judd, J. W. 1897. William Smith's manuscript maps. *Geological Magazine.* 34:439–47.

Kavaloski, Vincent. 1974. The vera causa principle: An historico-philosophical study of a metatheoretical concept from Newton to Darwin. Ph.D. diss., Univ. of Chicago.

Keir, James. 1776. On the crystallisations observed in glass. *Philosophical Transactions of the Royal Society of London.* 66:209–304.

Kennedy, Robert. 1805. A chemical analysis of three species of whinstone . . . and two of lava. *Transactions of the Royal Society of Edinburgh.* 5:76–98.

Kidd, John. 1809. *Outlines of mineralogy.* Oxford and London: J. Parker.

Kielmayr, Karl Friedrich. 1793. *Über die Verhältnisse der organischen Kräfte.* Stuttgart.

Kircher, Athanasius. 1665. *Mundus subterraneus.* Amsterdam.

Kirwan, Richard. 1793. Examination of the supposed igneous origin of stony substances. *Transactions of the Royal Irish Academy.* 5:51–81.

———. 1799. *Geological essays.* London.

———. 1800. *Observations on the proofs of the Huttonian theory of the earth adduced by Sir James Hall, Bart.* Dublin.

———. 1802a. An illustration and confirmation of some facts mentioned in an essay on the primitive state of the globe. *Philosophical Magazine.* 14:14–17.

———. 1802b. A reply to Mr. Playfair's reflections on Mr. Kirwan's refutation of the Huttonian theory of the earth. *Philosophical Magazine.* 14:3–13.

Kitts, David B. 1977. *The structure of geology.* Dallas: Southern Methodist Univ. Press.

Klaproth, Martin Heinrich. 1795–1815. *Beiträge zur chemischen Kenntniss der Mineralkörper.* 6 vols. Posen.

———. 1801. *Analytical essays towards promoting the chemical knowledge of mineral substances.* London: Cadell.

Kobell, Franz von. 1863. *Geschichte der Mineralogie von 1650–1850.* Munich: Merhoff.

Kubrin, David. 1967. Newton and the cyclical cosmos: Providence and the mechanical philosophy. *Journal of the History of Ideas.* 28:325–46.

———. 1968. Providence and the mechanical philosophy: The creation and dissolution of the world in Newtonian thought. Ph.D. diss., Cornell Univ.

Kuhn, Thomas. 1962. *The structure of scientific revolutions.* Chicago: Univ. of Chicago Press.

Lakatos, Imre. 1978. *The methodology of scientific research programmes.* Cambridge: Cambridge Univ. Press.

Lamarck, J. B. P. A. de Monet de. 1809. *Mémoires sur les fossiles des environs de Paris.* Paris. [Originally published as separate articles in *Annales du Muséum National d'Histoire Naturelle,* Paris, 1802–6.)]

Lamétherie, Jean-Claude de. 1797. 2d ed. *Théorie de la terre.* 5 vols. Paris.

Laplace, Pierre Simon Marquis de. 1824. *Exposition du système du monde*. 5th ed. Paris: Bachelier Janvier.

Larson, James L. 1971. *Reason and experience: The representation of natural order in the work of Carl von Linné*. Berkeley: Univ. of California Press.

Laudan, Larry. 1977. *Progress and its problems*. Berkeley: Univ. of California Press.

————. 1981. *Science and hypothesis: historical essays on scientific methodology*. Dordrecht, Netherlands: Reidel.

Laudan, Larry; A. Donovan; R. Laudan; P. Barker; H. Brown; J. Leplin; P. Thagard; and S. Wykstra. 1986. Scientific Change: Philosophical Models and Historical Research. *Synthese*. 69:141–223.

Laudan, Rachel [Bush]. 1974. The development of geological mapping in Britain from 1795 to 1825. Ph.D. diss., University of London.

Laudan, Rachel. 1976. William Smith: Stratigraphy without palaeontology. *Centaurus*. 20:210–26.

————. 1977a. Consolidation in the Huttonian tradition. *Lychnos*.195–206.

————. 1977b. Ideas and organizations: The case of the Geological Society of London. *Isis*. 68:527–38.

————. 1982. The role of methodology in Lyell's geology. *Studies in History and Philosophy of Science*. 13:215–50.

————. 1982b. Tensions in the concept of geology: Natural history or natural philosophy? *Journal of the History of the Earth Sciences*. 1:7–13.

————. 1983. Redefinitions of a discipline: Histories of geology and geological history, In *Functions and uses of disciplinary history*, eds. L. Graham, W. Lepenies, and P. Weingart, 79–104. Dordrecht, Netherlands: Reidel.

Launay, Louis de. 1779. *Mémoire sur l'origine des fossiles accidentales des provinces Belgique*. Brussels.

Lavoisier, Antoine Laurent. 1862. *Oeuvres de Lavoisier*. Paris: Imprimerie Impériale.

Lawrence, Philip. 1977. Heaven and earth—the relation of the nebular hypothesis to geology. In *Cosmology, history and theology*, ed. W. Yourgrau and D. Breck, 253–81. New York: Plenum.

————. 1978. Charles Lyell versus the theory of central heat: A reappraisal of Lyell's place in the history of geology. *Journal of the History of Biology*. 11:101–28.

Lehmann, Johann Gottlob. 1756. *Versuch einer Geschichte von Flötz-Gebürgen, betreffend deren Entstehung, Lage, darinne befindliche Metallen, Mineralien und Fossilien*. Berlin.

Lemaine, Gerard, et al. 1976. *Perspectives on the emergence of scientific disciplines*. The Hague: Mouton.

Lenoble, Robert. 1954. *La géologie au milieu de XVIIe siècle*. Paris: Université de Paris.

LeNoir, Timothy. 1981. The Göttingen school and the development of transcendental Naturphilosophie in the romantic era. *Studies in the History of Biology*. 5:111–205.

———. 1982. *The strategy of life: Teleology and mechanics in nineteenth century German biology*. Dordrecht, Netherlands: Reidel.

Leonhardt, Karl. 1823. *Charakteristik der Felsarten*. Heidelberg: J. Engelmann.

Leopold, Luna B., and Walter B. Langbein. 1963. Association and indeterminacy in geomorphology. In *The fabric of geology*, ed. Claude C. Albritton. Stanford, Calif.: Freeman.

Lindroth, Sten. 1983. The two faces of Linnaeus. In *Linnaeus: The man and his work*, ed. Tore Frängsmayr. Berkeley: Univ. of California Press.

Linnaeus, Carl. [1735] 1802. *A general system of nature, through the three grand kingdoms of animals, vegetables, and minerals*. Trans. W. Turton. London: Lackington Allen.

Linnaeus, Carl. 1751. *Philosophica botanica*. Stockholm.

Losee, John. 1972. *A historical introduction to the philosophy of science*. Oxford: Oxford Univ. Press.

Lovejoy, Arthur O. [1936] 1960. *The great chain of being*. Reprint. New York: Harper Torchbooks.

Lyell, Charles. 1825. On a dike of serpentine cutting through sandstone in the county of Forfar. *Edinburgh Journal of Science*. 3:112–26.

———. 1826. Review of *Transactions of the Geological Society of London, 1824*. *Quarterly Review*. 34:507–40.

———. 1827. Review of *Memoir on the geology of central France* by G. P. Scrope, F.R.S. *Quarterly Review*. 36:437–83.

———. 1830–33. *Principles of geology*. 3 vols. London: J. Murray.

———. 1847. *Principles of geology*. 7th ed. 3 vols. London: J. Murray.

———. 1850. On craters of deundation, with observations on the structure and growth of volcanic cones. *Quarterly Journal of the Geological Society of London*. 6:207–34.

———. 1858. On the structure of lavas which have consolidated on steep slopes, with remarks on the mode of origin of Mount Etna and on the theory of "craters of elevation." *Philosophical Transactions of the Royal Society*. 148:703–86.

Lyell, Katherine M. 1881. 2 vols. *Life, letters and journals of Sir Charles Lyell, Bart*. London: J. Murray.

Macculloch, John. 1819. *A description of the Western Isles of Scotland*. London.

———. 1821. *A geological classification of rocks*. London: Longman.

———. 1823. On certain elevations of land, connected with the actions of volcanoes. *Quarterly Journal of Science*. 14:262–95.

MacIntyre, D. B. 1963. James Hutton and the philosophy of geology. In *The fabric of geology*, ed. C. C. Albritton, 1-11. Stanford: Stanford University Press.

Maclure, William. 1809. Observations on the geology of the United States, explanatory of a geological map. *Transactions of the American Philosophical Society* 6:411–28.

Macquer, Pierre Joseph. [1766] 1777. *A dictionary of chemistry.* Trans. James Keir, 3 vols., 2d London edition. London: Cadell.

Marx, C. M. 1825. *Geschichte der Kristallkunde.* Karlsruhe.

Mathe, Gerhard. 1974. Leopold von Buch und seine Bedeutung für die Entwicklung der Geologie. *Zeitschrift für geologische Wissenschaft.* 2:1395–1404.

Mather, Kirtley F., and Shirley L. Mason. [1939] 1967. *A source book in geology: 1400–1900.* Reprint. Cambridge, Mass.: Harvard Univ. Press.

Mauskopf, Seymour H. 1970. Minerals, molecules and species. *Archives internationales d'histoire des sciences.* 23:185–206.

———. 1976. Crystals and compounds: Molecular structure and composition in nineteenth-century French science. *Transactions of the American Philosophical Society.* 66:1–82.

Mayr, Ernst. 1982. *The growth of biological thought: Diversity, inheritance and evolution.* Cambridge, Mass.: Harvard Univ. Press.

McCartney, Paul J. 1976. Charles Lyell and G. B. Brocchi: A study in comparative historiography. *British Journal for the History of Science.* 9:175–189.

McPhee, John. 1980. *Basin and range.* New York: Farrar, Straus and Giroux.

Melhado, Evan. 1980. Mitscherlich's discovery of isomorphism. *Historical Studies in the Physical Sciences.* 11:87–123.

———. 1981. *Jacob Berzelius: The emergence of his chemical system.* Stockholm: Almqvist and Wiksell.

Metzger, Helene. 1918. *La genèse de la science des cristaux.* Paris.

———. 1930. *Newton, Stahl, Boerhaave et la doctrine chimique.* Paris: Blanchard.

Michell, John. 1760. *Conjectures concerning the cause, and observations upon the phaenomena, of earthquakes.* London.

Mitscherlich, E. 1818–19. Über die Krystallisation der Salze . . . . *Abhandlungen der Königlichen Akademie der Wissenschaften, Berlin.* 5:427–37.

Mohs, Friedrich. 1820. *The characters of the classes, orders, genera, and species.* Edinburgh: Tait.

———. 1825. *Treatise on mineralogy; or, the natural history of the mineral kingdom.* Edinburgh: Constable.

Momigliano, Arnaldo. 1950. Ancient history and the antiquarian. *Journal of the Warburg and Courtauld Institutes.* 13:285–315.

Morello, Nicoletta. 1981. De Glossopetris Dissertatio: The demonstration by Fabio Colonna of the true nature of fossils. *Archives Internationales d'Histoire des Sciences.* 31:63–71.

Mornet, Daniel. 1911. *Les sciences de la nature en France, au XVIII siècle.* Paris: Colin.

Morrell, J. B. 1976. London institutions and Lyell's career: 1820–41. *British Journal for the History of Science*. 9:132–46.

Multhauf, Robert. 1958. The beginning of mineralogical chemistry. *Isis*. 49:50–53.

Murchison, Roderick. 1829. On the relations of the tertiary and secondary rocks forming the southern flanks of the Tyrolese Alps near Bassano. *Philosophical Magazine*. new ser. 5:401–10.

Murray, John. 1802. *Comparative view of the Huttonian and Neptunian systems*. Edinburgh: Ross and Blackwood.

———. 1815. On the diffusion of heat at the surface of the earth. *Transactions of the Royal Society of Edinburgh*. 7:411–34.

Nagel, Ernst. 1961. *The structure of science*. New York: Harcourt, Brace and World.

Newcomb, Sally. 1985. Laboratory evidence of silica solution supporting Wernerian theory. Paper presented at the 17th International Congress of History of Science, Berkeley, California.

Numbers, Ronald C. 1977. *Creation by natural law: Laplace's nebulae hypothesis in American thought*. Seattle: Univ. of Washington Press.

Oldroyd, David. 1970. The doctrine of property-conferring principles in chemistry: Origins and antecedents. *Organon*. 12/13:139–55.

———. 1971. The vulcanist-neptunist debate reconsidered. *Journal of Geological Education*. 19:124–29.

———. 1972. Robert Hooke's methodology of science as exemplified in his discourse of earthquakes. *British Journal for the History of Science*. 6:109–30.

———. 1973a. Some early usages of chemical terms. *Journal of Chemical Education*. 50:450–54.

———. 1973b. An examination of G. E. Stahl's *Philosophical principles of universal chemistry*. *Annals of Science*. 20:36–52.

———. 1973c. Some eighteenth century methods for the chemical analysis of minerals. *Journal of Chemical Education*. 50:337–40.

———. 1974a. Some phlogistic mineralogical schemes, illustrative of the evolution of the concept of "earth" in the seventeenth and eighteenth centuries. *Annals of Science*. 31:269–305.

———. 1974b. A note on the status of A. F. Cronstedt's simple earths and his analytical methods. *Isis*. 65:506–12.

———. 1974c. Mechanical mineralogy. *Ambix*. 21:157–78.

———. 1974d. From Paracelsus to Haüy: The development of mineralogy in its relation to chemistry. Ph.D. diss., Univ. of New South Wales.

———. 1974e. Some Neoplatonic and Stoic influences on mineralogy in the sixteenth and seventeenth centuries. *Ambix*. 21:128–56.

———. 1975. Mineralogy and the "Chemical Revolution." *Centaurus*. 1:54–71.

———. 1979. Historicism and the rise of historical geology. *History of Science*. 17:191–213.

Olson, Richard. 1975. Scottish philosophy and British physics, 1750–1880. Princeton, N.J.: Princeton Univ. Press.

Omalius d'Halloy, J. B. J. d'. 1822. Observations sur un essai de carte géologique de la France, des Pays-Bas, et des contrées voisins. *Annales des Mines.* 7:353–76. Trans. De la Beche (1824). *A selection of geological memoirs.* London: Phillips.

———. 1823. "Observations sur un essai de carte géologique de la France." *Annales des Mines* 7:353–76.

O'Rourke, J. E. 1978. A comparison of James Hutton's *Principles of knowledge* and *Theory of the earth. Isis.* 69:5–20.

Ospovat, Alexander M. 1967. The place of the *Kurze Klassifikation* in the work of A. G. Werner. *Isis.* 58:90–95.

———. 1969. Reflections on A. G. Werner's "Kurze Klassifikation." In *Toward a history of geology,* ed. C. Schneer, 242–56. Cambridge, Mass.: MIT Press.

———. 1971. Introduction and Notes. In edition of A. G. Werner, *Short classification and description of the various rocks.* New York: Hafner.

———. 1976. The distortion of Werner in Lyell's *Principles of Geology. British Journal for the History of Science.* 9:190–98.

———. 1979. Werner's concept of the basement complex. In Geological Association of Canada Special Paper 19, ed. W. O. Kupsch and W. A. S. Sargeant, 161–70.

———. 1980. The importance of regional geology in the geological theories of Abraham Gottlob Werner: A contrary opinion. *Annals of Science.* 37:433–40.

———. 1982. Romanticism and geology: Five students of Abraham Gottlob Werner. *Eighteenth Century Life.* 7:105–17.

Ospovat, Dov. 1977. Lyell's theory of climate. *Journal of the History of Biology.* 10:317–39.

Outram, Dorinda. 1978. The language of natural power: The éloges of Georges Cuvier and the public language of nineteenth-century science. *History of Science.* 16:153–78.

Page, Leroy E. 1969. Diluvialism and its critics in Great Britain in the early nineteenth century. In *Toward a history of geology,* ed. C. Schneer, 257–71. Cambridge, Mass.: MIT Press.

Parkinson, James. 1804–11. *Organic remains of a former world: An examination of the mineralized remains of the vegetables and animals of the antediluvian world; generally termed extraneous fossils.* 3 vols. London: Robson.

———. 1811. Observations on some of the strata in the neighbourhood of London, and on the fossil remains contained in them. *Transactions of the Geological Society of London.* 1:324–54.

Partington, J. R. 1961–64. *A history of chemistry.* London.

Pearce Williams, L., ed. 1971. *The selected correspondence of Michael Faraday.* Cambridge: Cambridge Univ. Press.

Phillips, John. 1844. *Memoirs of William Smith, L.L.D.* London: Murray.

Phillips, William. 1816. *Outline of mineralogy and geology.* London: Phillips.

Playfair, John. [1802] 1956. *Illustrations of the Huttonian theory of the earth.* Reprint, with introduction by G. White, Urbana: Univ. of Illinois Press.

―――. 1805. Biographical account of the late Dr. James Hutton. *Transactions of the Royal Society of Edinburgh.* 5:39–99.

―――. 1822. *Works.* 4 vols. Edinburgh: Constable.

Pollard, Sidney. 1965. *The genesis of modern management.* London: Edward Arnold.

―――. 1981. *Peaceful conquest.* Oxford: Oxford Univ. Press.

Porter, Roy. 1973. The industrial revolution and the emergence of the science of geology. In *Changing perspectives in the history of science*, ed. M. Teich and R. Young, 320–43. London: Heinemann.

―――. 1976. Charles Lyell and the principles of the history of geology. *British Journal for the History of Science.* 9:91–103.

―――. 1977. *The making of geology: Earth science in Britain, 1660–1815.* Cambridge: Cambridge Univ. Press.

―――. 1980. The terraqueous globe. In *The ferment of knowledge*, ed. G. S. Rousseau and R. Porter, 285–326. Cambridge: Cambridge Univ. Press.

Porter, Roy, and K. Poulton. 1977. Research in British geology 1660–1800: A survey and thematic bibliography. *Annals of Science.* 34:33–42.

―――. 1978. Geology in Britain, 1660–1800: A selective biographical bibliography. *Journal of the Society for the Bibliography of Natural History.* 9:74–84.

Porter, Theodore M. 1981. The promotion of mining and the advancement of science: The chemical revolution and mineralogy. *Annals of Science.* 38:543–70.

Pott, Johann Heinrich. 1746. *Lithogeognosia: Chymsiche Untersuchungen.* Potsdam.

Prescher, Hans, et al. 1980. Johannes Kentmanns Mineralienkatalog aus dem Jahre 1565. *Abhandlungen des Staatlichen Museums für Mineralogie und Geologie zu Dresden.* 30:21–35.

Prévost, Constant. 1823. De l'importance de l'étude des corps organisés vivants pour la géologie positive. *Mémoires de la Societé d'Histoire Naturelle de Paris.* 1:259–68.

―――. 1835. Notes sur l'île Julia, pour servir à l'histoire de la formation des montagnes volcaniques, *Mémoires de la Societé géologique de France.* 2:91–124.

Prevost, Pierre. 1792. *Recherches physico-mécaniques sur la chaleur.* Geneva: Barde, Manget & Co.

Ramsay, William. 1918. *The life and letters of Joseph Black, M.D.* London: Constable.

Rappaport, Rhoda. 1960. G.-F. Rouelle: An eighteenth-century chemist and teacher. *Chymia.* 6:68–101.

————. 1964. Problems and sources in the history of geology. *History of Science.* 3:60–78.

————. 1968. Lavoisier's geologic activities, 1763–1792. *Isis.* 58:375–84.

————. 1969a. The early disputes between Lavoisier and Monnet, 1777–1781. *British Journal for the History of Science.* 4:233–44.

————. 1969b. The geological atlas of Guettard, Lavoisier, and Monnet: Conflicting views of the nature of geology. In *Toward a history of geology,* ed. C. Schneer, 272–87. Cambridge, Mass.: MIT Press.

————. 1971. Commentary on the paper of M. J. S. Rudwick. In *Perspectives in the history of science and technology,* ed. Duane Roller. 228–31. Norman: Univ. of Oklahoma Press.

————. 1973. Lavoisier's theory of the earth. *British Journal for the History of Science.* 6:247–60.

————. 1978. Geology and orthodoxy: The case of Noah's flood in eighteenth-century thought. *British Journal for the History of Science.* 11:1–18.

————. 1982. Borrowed words: Problems of vocabulary in eighteenth century geology. *British Journal for the History of Science.* 15:27–44.

Raspe, Rudolf Erich. [1763] 1970. *An introduction to the natural history of the terrestrial sphere.* Trans. and ed. A. N. Iversen and A. V. Carozzi. New York: Hafner.

————. 1776. *An account of some German volcanos, and their productions.* London: L. Davis.

Ray, John. 1692. *Miscellaneous discourses concerning the dissolution and changes of the world.* London: S. Smith.

Réaumur, R. A. F. de. 1723. Sur la nature et la formation des cailloux. *Mémoires de l'Académie Royale des Sciences, 1721.* 255–76.

————. 1725. Sur la rondeur que semblent affecter certaines espèces de pierres. *Mémoires de l'Académie Royale des Sciences, 1723.* 273–84.

————. 1727. Idée genérale des différentes manières dont on peut faire la porcelaine, et quelles sont les veritables matières de celles de Chine. *Mémoires de l'Académie Royale des Sciences, 1727.* 185–203.

————. 1729. Second mémoire sur la porcelaine, on suite des principes qui doivent conduire dans la composition de porcelaines de differens genres. *Mémoires de l'Académie Royale des Sciences, 1729.* 325–44.

————. 1732. De la nature de la terre en general, et du caractère des différentes espèces. *Mémoires de l'Académie Royale des Sciences, 1730.* 243–83.

Rehbock, Philip. 1983. *The philosophical naturalists: Themes in early nineteenth-century British biology.* Madison: Univ. of Wisconsin Press.

Reid, Thomas. [1785] 1969. *Essay on the intellectual powers of man.* Reprint of 1813–15 collected edition. Cambridge, Mass.: MIT Press.

Reuss, Franz Ambrosius. 1801–6. *Lehrbuch der Mineralogie nach des Herrn D. L. G. Karsten mineralogischen Tabellen.* 8 vols. Leipzig: F. G. Jacobder.

Review of Hutton. 1788. *Analytical Review.* 1:424–25.

Review of Hutton. 1788. *Critical Review.* 66:115–20.

Review of the "Illustrations of the Huttonian Theory of the Earth." 1802. *Edinburgh Review.* 1:211.

Richards, Robert. 1979. Influence of the sensationalist tradition on early theories of the evolution of behavior. *Journal of the History of Ideas.* 11:86–105.

Richardson, William. 1808. A letter on the alterations that have taken place in the structure of the rocks, on the surface of the basaltic country in the counties of Derry and Antrim. *Philosophical Transactions of the Royal Society of London.* 98:187–222.

Robinet, Jean-Baptiste. 1761–66. *De la nature.* 4 vols. Paris.

Rocque, Aurele La. 1969. Bernard Palissy. In *Toward a history of geology*, ed. C. Schneer, 226–41. Cambridge, Mass.: MIT Press.

Roger, Jacques. 1962. "Buffon: *Les Époques de la nature*, edition critique. *Mémoires du Muséum Nationale d'Histoire Naturelle.* ser. c. 10. Paris.

———. 1973. La théorie de la terre au XVII siècle. *Revue d'Histoire des Sciences.* 26:23–48.

Rudwick, Martin. 1962a. Hutton and Werner compared. George Greenough's geological tour of Scotland in 1805. *British Journal for the History of Science.* 1:117–35.

———. 1962b. The principle of uniformity. *History of Science.* 2:82–86.

———. 1963. The foundation of the Geological Society of London: Its scheme for co-operative research and its struggle for independence. *British Journal for the History of Science.* 1:324–55.

———. 1964. The inference of function from structure in fossils, *British Journal for the Philosophy of Science.* 15:27–40.

———. 1967. A critique of uniformitarian geology: A letter from W. D. Conybeare to Charles Lyell, 1841. *Proceedings of the American Philosophical Society.* 111:272–87.

———. 1969. Lyell on Etna, and the antiquity of the earth. In *Toward a history of geology*, ed. Cecil J. Schneer. 288–304. Cambridge, Mass.: MIT Press.

———. 1970. The strategy of Lyell's *Principles of Geology. Isis.* 61:5–33.

———. 1971. Uniformity and progression: Reflections on the structure of geological theory in the age of Lyell. In *Perspectives in the history of science and technology*, ed. D. H. D. Roller, 209–27. Norman, Oklahoma: Univ. of Oklahoma Press.

———. 1972. *The meaning of fossils: Episodes in the history of palaeontology.* London: Macdonald.

———. 1974a. Darwin and Glen Roy: A "great failure" in scientific method? *Studies in History and Philosophy of Science.* 5:97–185.

———. 1974b. Poulett Scrope on the volcanos of Auvergne: Lyellian time and political economy. *British Journal for the History of Science.* 7:205–42.

————. 1976. The emergence of a visual language for geological science. *History of Science*. 14:149–95.

————. 1978. Charles Lyell's dream of a statistical palaeontology. *Palaeontology*. 21:225–44.

————. 1985. *The great Devonian controversy*. Chicago: Univ. of Chicago Press.

Rupke, Nicholaas. 1983a. The study of fossils in the romantic philosophy of history and nature. *History of Science*. 21:389–413.

————. 1983b. *The great chain of history: William Buckland and the English school of geology (1814–1849)*. Oxford: Clarendon.

Ruse, Michael. 1976. Charles Lyell and the philosophers of science. *British Journal for the History of Science*. 9:121–31.

————. 1979. *The Darwinian revolution: Science red in tooth and claw*. Chicago: Univ. of Chicago Press.

Sainte-Claire Deville, Charles. 1878. *Coup-d'oeuil historique sur la géologie et sur les travaux d'Élie de Beaumont*. Paris.

Sarjeant, W. A. S. 1980. *Geologists and the history of geology: An international bibliography from the origins to 1978*. New York and London: Arno Press and Macmillan.

Saussure, Horace Bénédict de. 1779–96. *Voyages dans les Alpes, precédés d'un essai sur l'histoire naturelle des environs de Genève*. Neuchatel: S. Fauche.

Scherz, Gustav. 1963. *Ein Pioneer der Wissenschaft: Niels Stensen in seinen Schriften*. Copenhagen: Munksgaard.

————, ed. 1971. *Dissertations on Steno as a geologist*. Odense: Odense Univ. Press.

Schlotheim, E. F. von. 1801a. Beiträge zur nähern Kenntniss einzelner Fossilien. *Magazin für die gesammte Mineralogie, Geognosie und Mineralogische Erdbeschreibung*. 1:143–72.

————. 1801b. Über die Kräuter-Abdrucke in Schieferthon und Sandstein. . . . *Magazin für die gesammte Mineralogie, Geognosie und Mineralogische Erdbeschreibung*. 1:79–95.

————. 1804. *Beschreibung merkwurdiger Kräuter-Abdrucke und Pflanzen-Versteinerungen*. Gotha.

————. 1813. Beiträge zur Naturgeschichte der Versteinerungen in geognostischer Hinsicht. *Taschenbuch für die gesammte Mineralogie*. 7:3–134. Partially translated in *A source book in geology*, ed. K. Mather and S. Mason (1967), 174–75. Cambridge, Mass.: Harvard Univ Press.

————. 1822. *Nachtrage zur Petrefactenkunde*. Gotha: Becker.

Schneer, Cecil. 1954. The rise of historical geology in the seventeenth century. *Isis*. 45:256–68.

————, ed. 1969. *Toward a history of geology*. Cambridge, Mass.: MIT Press.

————. 1981. Aspects of form and structure: The renaissance background to crystallography," in Harry Woolf, (ed.), *The Analytic Spirit*: Ithaca, N.Y.: Cornell Univ. Press.

Schofield, Robert E. 1970. *Mechanism and materialism.* Princeton, N.J.: Princeton Univ. Press.

Schufle, J. A. 1967. Torbern Bergman, earth scientist. *Chymia.* 12:59–97.

Scrope, George Poulett. 1825. *Considerations on volcanos: The probable causes of their phenomena, the laws which determine their march, the disposition of their products.* London: Phillips.

———. 1827. *Memoir on the geology of central France; including the volcanic formations of Auvergne, the Velay, and the Vivarais.* London: Longman.

———. 1830. *Principles of Geology* . . . by Charles Lyell . . . vol. I. *Quarterly Review.* 43:293–99.

———. 1858. *The geology and extinct volcanos of central France.* London: J. Murray.

———. 1862. *Volcanos: The character of their phenomena, their share in the structure and composition of the surface of the globe, and their relation to its internal forces.* London: Longman, Green, Longmans, and Roberts.

Sedgwick, Adam. 1822. On the geology of the Isle of Wight. *Annals of Philosophy.* 2 ser. 3:329–55.

———. 1827. On the association of trap rocks with the mountain limestone formation in High Teesdale. [1823–24] *Transactions of the Cambridge Philosophical Society.* 2:139–96.

———. 1833a. On the magnesian limestone of the northern counties. [1826] *Proceedings of the Geological Society of London.* 1:21–22.

———. 1833b. On some beds associated with the magnesian limestone, and on some fossil fish found in them. [1826] *Proceedings of the Geological Society of London.* 1:2–3.

———. 1833c. Presidential address to the Geological Society. *Proceedings of the Geological Society of London.* 1:281–336.

———. 1833d. On the geological relations and internal structure of the magnesian limestone, and the lower portions of the New Red Sandstone series in their range through Nottinghamshire, Derbyshire, Yorkshire and Durham, to the southern extremity of Northumberland. [1826–28]. *Proceedings of the Geological Society of London.* 1:63–66.

———. 1835. Remarks on the structure of large mineral masses, and especially on the chemical changes produced in the aggregation of stratified rocks during different periods after their deposition. *Transactions of the Geological Society of London.* 2d ser. 3:461–86.

Sedgwick, Adam, and Murchison, R. 1830. A sketch of the structure of the Austrian Alps. *Philosophical Magazine.* new ser. 8:80–134.

Shapin, S. 1974. The audience for science in eighteenth-century Edinburgh. *History of Science.* 12:95–121.

Sherley, Thomas. 1672. *A philosophical essay declaring the probable causes, whence stones are produced in the greater world.* London.

Siegfried, Robert, and Dobbs, Betty Jo. 1968. Composition: A neglected aspect of the chemical revolution. *Annals of Science.* 24:275–93.

/

Siegfried, Robert, and Dott, Robert H. 1976. Humphry Davy as geologist, 1805–29. *British Journal for the History of Science*. 9:219–27.

———. 1980. *Humphry Davy on geology: The 1805 lectures for the general audience*. Madison: Univ. of Wisconsin Press.

Simpson, George Gaylord. 1963. Historical science. In *The fabric of geology*, ed. Claude C. Albritton. Stanford, Calif.: Freeman.

Sloan, Philip R. 1972. John Locke, John Ray, and the problem of the natural system. *Journal of the History of Biology*. 5:1–53.

———. 1979. Buffon, German biology, and the historical interpretation of biological species. *British Journal for the History of Science*. 12:109–53.

Smith, Cyril Stanley. 1960. *A history of metallography*. Chicago: Univ. of Chicago Press.

———. 1969. Porcelain and plutonism. In *Toward a history of geology*, ed. C. Schneer, 317–38. Cambridge, Mass.: MIT Press.

Smith, W. Campbell. 1978. Early mineralogy in Great Britain and Ireland. *Bulletin of the British Museum (Natural History)*. 6:49–74.

Smith, William 1815a. *A delineation of the strata of England and Wales, with part of Scotland*. London.

———. 1815b. *A memoir to the map*. London.

———. 1816–19. *Strata identified by organized fossils*. London.

———. 1817. *Stratigraphical system of organized fossils*. London.

Sowerby, James. 1812. *The mineral conchology of Great Britain*. London: B. Meredith.

St. Clair, Charles. 1966. The classification of minerals: Some representative mineral systems from Agricola to Werner. Ph.D. diss., University of Oklahoma.

Steffens, Heinrich. 1810. *Geognostisch-geologische Aufsätze als Vorbereitung zu einer innern Naturgeschichte der Erde*. Hamburg: Hoffmann.

———. 1811. *Vollständiges Handbuch der Oryktognosie*. Halle: Curts.

Steno, Nicolaus. [1671] 1968. *The prodromus to a dissertation concerning solids naturally contained within solids*. Trans. John Winter (1918), Reprint. New York: Hafner.

Stewart, Dugald. 1792. *Philosophy of the human mind*. Edinburgh.

Stoddart, D. R. 1976. Darwin, Lyell, and the geological significance of coral reefs. *British Journal for History of Science*. 9:199–218.

Strachey, John. 1725. An account of the strata in coal-mines, etc. *Philosophical Transactions of the Royal Society of London*. 33:395–98.

Studer, Bernhard. 1834. *Geologie der westlichen Schweizer-Alpen: Ein Versuch*. Heidelberg: K. Gross.

Stukeley, William. 1750. *The philosophy of earthquakes, natural and religious*. London: C. Corbet.

Suess, Eduard. *Das Antlitz der Erde* [1885–1909] 1904–9. Trans. Hertha B. C. Sollas, under direction of W. J. Sollas. 4 vols. Oxford: Clarendon Press.

Sweet, Jessie M., and Charles D. Waterstown. 1967. Robert Jameson's

approach to the Wernerian theory of the earth, 1796. *Annals of Science.*
23:81–95.

Taton, Renée. 1964. *Enseignement et diffusion des sciences en France au XVIII siècle.* London: W. Blackwood.

Taylor, Kenneth L. 1968. Nicolas Desmarest (1725–1815): Scientist and industrial technologist. Ph.D. diss. Harvard Univ.

———. 1969. Nicolas Desmarest and geology in the eighteenth century. In *Toward a history of geology,* ed. Cecil Schneer, 339–56. Cambridge, Mass.: MIT Press.

———. 1975. "Changing conceptions of crystallization and vitrification: Their role in geology, 1750–1825." Paper presented at the 1975 meeting of the Geological Society of America, Salt Lake City.

———. 1977. The new chemistry and volcanology: Chemical theories of volcanic action, 1790–1830. Paper presented at the 1977 meeting of the History of Science Society, New York.

———. 1979. Geology in 1776: Some notes on the character of an incipient science. In *Two hundred years of geology in America,* ed. C. Schneer. Cambridge, Mass.: MIT Press.

———. 1981–82. The beginnings of a French geological identity. *Histoire et Nature.* 19–20:65–82.

———. 1985. Early geoscience mapping, 1700–1830. *Proceedings of the Geoscience Information Society.* 15:15–49.

Tennant, Smithson. 1799. On different sorts of lime used in agriculture. *Philosophical Transactions of the Royal Society of London.* 89:548–53.

Thackray, Arnold. 1970. *Atoms and powers.* Cambridge, Mass.: Harvard Univ. Press.

Theophrastus. 1956. *On stones.* Ed. and trans. R. R. Cayley and J. C. F. Richards. Columbus, Ohio: Ohio State Univ. Press.

Thomson, Thomas. 1813. Some observations in answer to Mr. Chevenix's attack upon Werner's mineralogical method. *Annals of Philosophy.* 1:241–58.

Todhunter, Isaac. [1873] 1962. *The figure of the earth.* 2 vols. Reprint. New York: Dover.

Tomkeiff, S. I. 1950. James Hutton and the philosophy of geology. *Proceedings of the Royal Society of Edinburgh.* 63:387–400.

Torrens, Hugh. 1978. Geological communication in the Bath area in the last half of the eighteenth century. In *Images of the earth: Essays in the history of the environmental sciences,* ed. L. Jordanova and R. Porter, 215–47. Chalfont St. Giles: British Society for the History of Science.

———. 1981. The Reynolds-Austice Shropshire geological collection—200 years of history and its lessons. In *History in the service of systematics.* London: Society for the Bibliography of Natural History.

———. 1984. The history of coal prospecting in Britain, 1650–1900. In *Energie in der Geschichte.* Dusseldorf: Verein Deutscher Ingenieure.

Toulmin, Stephen, and Goodfield, June. 1965. *The discovery of time.* New York: Harper & Row.

Turner, R. Steven. 1974. Reformers and scholarship in Germany. In *The university in society*, ed. Lawrence Stone, 2:495–531. Princeton, N.J.: Princeton Univ. Press.

Vallance, T. G., and H. S. Torrens. 1984. The Anglo-Austrian traveller Robert Townson and his map of Hungarian 'petrography' [1797]. In *Contributions to the history of geological mapping*, ed. E. Dudich. Budapest: Akademiai Kiado.

Vitruvius. 1914. *The ten books of architecture*. Trans. M. H. Hickey. Cambridge, Mass.: Harvard Univ. Press.

Vogelsang, H. 1867. *Philosophie der Geologie und mikroskopische Gesteinsstudien*. Bonn: Cohen.

Voigt, J. K. W. 1781–85. *Mineralogische Reisen durch das Herzogthum Weimar und Eisenach*. 2 vols. Weimar.

————. 1787. *Mineralogische Reise von Weimar über den Thüringer Wald und Meiningen bis Hanau*. Leipzig.

————. 1802. *Mineralogische Reise nach den Braunkohlenwerken und Basalten in Hessen*. Weimar.

Wagenbreth, Otfried. 1953a. Leopold von Buch zu seinem 100. Todestag. *Bergakademie*. 5:92–101.

————. 1953b. Leopold von Buch und die Entwicklung der Theorie über Gebirgsbildung und Vulkanismus. *Bergakademie*. 332–38, 369–74.

————. 1955. Abraham Gottlob Werner und der Hohepunkt des Neptunistenstreits um 1790. *Freiberger Forschungshefte*. D11:183–241.

————. 1967. Abraham Gottlob Werners System der Geologie, Petrographie und Lagerstätten Lehre; *and* Werner-Schüler als Geologen und Bergleute und ihre Bedeutung für die Geologie und den Bergbau des 19. Jahrhunderts. In *Abraham Gottlob Werner Gedenkenschrift*, 83–148, 163–78. Leipzig.

————. 1968. Die Paläontologie in Abraham Gottlob Werners Geologischen System. *Bergakademie*. 20:32–36.

————. 1980. Die Geologie A. G. Werners in ihrer Wirkung von der Aufklarung bis Heute. *Zeitschrift für Geologische Wissenschaften*. 8:79–86.

Wagenbreth, Otfried, and Wachtler, Eberhard. 1981. Zur Wechselwirkung regional geologischer und gesellschaftlicher Faktoren in der Geschichte der geologischen Wissenschaften. *Freiberger Forschungshefte*. D132.

Walch, Johann. 1755–73. *Die Naturgeschichte der Versteinerungen zur Erlauterung der Knorrischen Sammlung von Merkwurdigkeiten der Natur*. Nurnberg.

Walker, John. 1966. *Lectures on geology*. Chicago: Univ. of Chicago Press.

Wallerius, Johan Gottschalk. 1753. *Minéralogie; ou description genérale des substances du regne minéral*. Paris: Durand, Pissot.

Walsh, Francis. 1743. *The antediluvian world; or, A new theory of the earth*. Dublin: S. Powell.

Warren, Erasmus. [1690] 1978. *Geologia or a discourse concerning the earth before the deluge*. Reprint. New York: Arno Press.

Watson, Richard. 1969. Explanation and prediction in geology. *Journal of Geology*. 77:488–94.

Watznauer, Adolf. 1960. Alexander von Humboldt und der Freiberger Kreis. *Freiberger Forschungshefte*. D33.

Weber, W. 1976. Innovationen im frühindustriellen deutschen Bergbau und Hüttenwesen: Friedrich Anton von Heynitz. *Studien zu Naturwissenschaft, Technik und Wirtschaft in Neuzehnten Jahrhundert*. 6.

Webster, Charles. 1966. Water as the ultimate principle of nature: The background to Boyle's Sceptical Chymist. *Ambix*. 13:96–107.

Webster, John. 1671. *Metallographia: or, An history of metals*. London: W. Kettilby.

Webster, Thomas. 1814. On the freshwater formations of the Isle of Wight, with some observations on the strata over the Chalk in the south-east part of England. *Transactions of the Geological Society of London*. 2:161–254.

Wegmann, Eugene. 1969. Changing ideas about moving shorelines. In *Toward a history of geology*, Ed. C. Schneer, 386–414. Cambridge, Mass.: MIT Press.

Weindling, Paul. 1979. Geological controversy and its historiography: The prehistory of the Geological Society of London. In *Images of the earth: essays in the history of the environmental sciences*, Ed. L. Jordanova and R. S. Porter, 248–71. Chalfont St. Giles: British Society for the History of Science.

———. 1983. The British Mineralogical Society: A case study in science and social improvement. In *Metropolis and province: Science in British culture, 1780–1850*, Ed. Ian Inkster and Jack Morrell, 120–150. London: Hutchinson.

Werner, Abraham Gottlob. [1774] 1962. *On the external characters of minerals*. Trans. Albert Carozzi of *Von den äusserlichen Kennzeichen der Fossilien*. Urbana: Univ. of Illinois Press.

———. 1786. *Short classification and description of the various rocks*. Trans. with introduction by A. Ospovat of *Kurze Klassifikation und Beschreibung der verschiedenen Gebirgsarten*. New York: Hafner.

[———]. 1789a. Mineralsystem der Herrn Inspektor Werners mit dessen Erlaubnis herausgegeben von C. A. S. Hoffmann. *Bergmännisches Journal*. 1:369–98.

———. 1789b. Versuch einer Erklärung der Entstehung der Vulkanen durch die Entzündung mächtiger Steinkohlenschichten, als ein Beytrag zu der Naturgeschichte des Basalts. *Magazin für die Naturgeschichte Helvetiens*. 4:239–54.

———. 1791. *Neue Theorie von der Entstehung der Gänge, mit Anwendung auf den Bergbau*. Freiberg: Gerlach.

———. 1809. *New theory of the formation of veins; with its application to the art of working mines*. Edinburgh: A. Constable.

———. 1817. *Mineral-System*. Freyberg: Graz und Gerlach.

Whewell, William. 1831. *Principles of geology*, vol. I, by Charles Lyell. *British Critic*. 17:180–206.

———. 1832. *Principles of geology*, vol. II, by Charles Lyell. *Quarterly Review*. 47:103–22.

———. 1857. *History of the inductive sciences, from the earliest to the present times*. 1967 Reprint of 3d ed. (1st ed., 1837). London: Frank Cass. London: Parker.

———. 1840. *Philosophy of the inductive sciences*. London: Parker.

Whiston, William. 1696. *A new theory of the earth, from its original, to the consummation of all things*. London: B. Tooke.

Whitehurst, John. 1778. *An inquiry into the original state and formation of the earth*. London.

Williams, John. 1789. *Natural history of the mineral kingdom*. Edinburgh.

Wilson, J. Tuzo. 1954. The development and structure of the crust. In *The earth as a planet*, ed. Gerald Kuiper, 138–214. Chicago: Chicago Univ. Press.

Wilson, Leonard G. 1969. The intellectual background to Charles Lyell's *Principles of geology*, 1830–1833. In *Toward a history of geology*, ed. C. Schneer, 426–43. Cambridge, Mass.: MIT Press.

———. 1972. *Charles Lyell: The years to 1841—the revolution in geology*. New Haven, Conn.: Yale Univ. Press.

Woodward, H. B. 1909. *The history of the Geological Society of London*. London: Longmans.

Woodward, John. 1695. *An essay toward a natural history of the earth: and terrestrial bodies, especially minerals: as also of the sea, rivers, and springs*. London: R. Wilkin.

———. 1728. *Fossils of all kinds, digested into a method, suitable to their mutal relation and affinity*. London: W. Innys.

———. 1729. *An attempt towards a natural history of the fossils of England*. London: F. Fayram.

Wyckoff, Dorothy, ed. 1967. *Albertus Magnus' Book of Minerals*. Oxford: Oxford Univ. Press.

Yearly, Steven. 1980. Textual persuasion: The role of social accounting in the construction of scientific arguments. *Philosophy of the Social Sciences*. 11:409–35.

———. 1984. Proofs and reputations: Sir James Hall and the use of classification devices in scientific argument. *Earth Sciences History*. 3:25–43.

Zittel, Karl A. von. 1901. *History of geology and palaeontology to the end of the 19th century*. London: Walter Scott.

Index